What's Happening in the Mathematical Sciences

Volume 10

What's Happening in the Mathematical Sciences

2010 *Mathematics Subject Classification*: 00A06

ISBN-13: 978-1-4704-2204-2
ISBN-10: 1-4704-2204-2

For additional information and updates on this book, visit
www.ams.org/bookpages/happening-10

Printed in the United States of America.
♾ The paper used in this book is acid-free and falls within the guidelines established to ensure permanence and durability.
Visit the AMS home page at
http://www.ams.org/

10 9 8 7 6 5 4 3 2 1 20 19 18 17 16 15

The cover for this publication was prepared using the Adobe® CS6® suite of software and the frontmatter was prepared using the Adobe® CC® software. The articles were prepared using TEX. TEX is a trademark of the American Mathematical Society.

About the Authors

DANA MACKENZIE is a mathematician who went rogue and turned into a freelance writer. He earned a Ph.D. in mathematics from Princeton University in 1983 and a graduate certificate in science communication from the University of California at Santa Cruz in 1997. Since then he has written for such magazines as *Science*, *New Scientist*, *Scientific American*, *Discover*, and *Smithsonian*. His most recent book, *The Universe in Zero Words* (Princeton University Press, 2012) is a history of 24 great equations in math and science. He received the George Pólya Award for exposition from the Mathematical Association of America in 1993, the Communication Award of the Joint Policy Board for Mathematics in 2012, and the Chauvenet Prize from the Mathematical Association of America in 2015.

BARRY CIPRA is a freelance mathematics writer based in Northfield, Minnesota. He received the 1991 Merten M. Hasse Prize from the Mathematical Association of America for an expository article on the Ising model, published in the December 1987 issue of the *American Mathematical Monthly*, and the 2005 Communication Award of the Joint Policy Board for Mathematics. His book, *Misteaks ... and how to find them before the teacher does...* (a calculus supplement), is published by AK Peters, Ltd.

About the Cover Images

Main image: Curved folds represent both a departure from tradition in origami and an area of active research from the artists, Erik Demaine and Martin Demaine. Below main image, left to right: A self-folding robot from Wyss Institute at Harvard University; a climate network developed at the Potsdam Institute for Climate Impact Research (PIK); Sherlock Holmes and John Watson, a line drawing by Sidney Paget, 1892; and coral in the Caribbean Sea overgrown by algae, an example of a climate "tipping point," from Craig Quirolo, Reef Relief. Back cover image: If each corner in the above path represents a city on a map, then this path is the shortest closed curve joining all the cities. Thus it solves the so-called "Traveling Salesman Problem" (TSP) for this particular map, from Robert Bosch.

Contents

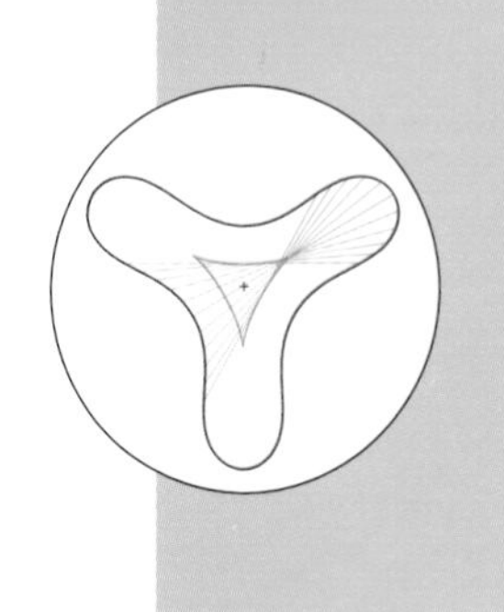

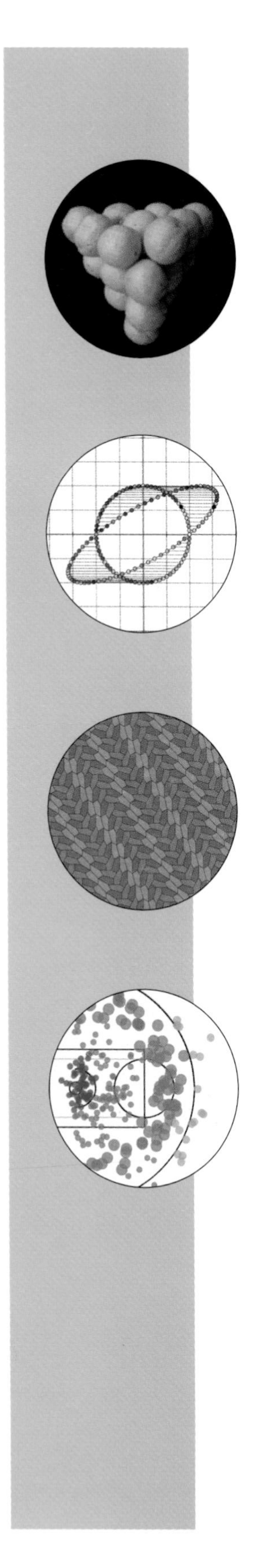

Introduction

SEVERAL IMPORTANT RECENT DEVELOPMENTS in pure mathematics are featured in this volume of *What's Happening in the Mathematical Sciences*. **"Prime Clusters and Gaps: Out-Experting the Experts,"** page 18, talks about new insights into the distribution of prime numbers, the perpetual source of new problems, and new results. Recently, several mathematicians (including Yitang Zhang and James Maynard) significantly improved our knowledge of the distribution of prime numbers. Advances in the so-called Kadison-Singer problem and its applications in signal processing algorithms used to analyze and synthesize signals are described in **"The Kadison-Singer Problem: A Fine Balance,"** page 72. **"Quod Erat Demonstrandum,"** page 64, presents two examples of perseverance in mathematicians' pursuit of truth using, in particular, computers to verify their arguments. Also, **"Following in Sherlock Holmes' Bike Tracks,"** page 52, shows how an episode in one of Sir Arthur Conan Doyle's stories about Sherlock Holmes naturally led to very interesting problems in the theory of completely integrable systems.

On the applied side, **"Climate Past, Present, and Future,"** page 36, shows the importance of mathematics in the study of problems of climate change and global warming. Mathematical models help researchers to understand the past, present, and future changes of climate, and to analyze their consequences. Economists have known for a long time that trust is a cornerstone of commerce. **"The Truth Shall Set Your Fee,"** page 28, shows how recent advances in theoretical computer science led to the development of so-called "rational protocols" for information exchange, where the seller of information is forced to tell the truth in order to maximize profit.

Over the last 100 years many professional mathematicians and devoted amateurs contributed to the problem of finding polygons that can tile the plane, e.g., used as floor tiles in large rooms and walls. Despite all of these efforts, the search is not yet complete, as the very recent discovery of a new plane-tiling pentagon shows in **"A Pentagonal Search Pays Off,"** page 86. The increased ability to collect and process statistics, big data, or "analytics" has completely changed the world of sports analytics as shown in **"The Brave New World of Sports Analytics,"** page 96. The use of modern methods of statistical modeling allows coaches and players to create much more detailed game plans in professional baseball and basketball as well as create many new ways of measuring a player's value. Finally, **"Origami: Unfolding the Future,"** page 2, talks about the ancient Japanese paper-folding art and origami's unexpected connections to a variety of areas including mathematics, technology, and education.

Sergei Gelfand, Publisher

Fire Tower (2013). *Curved folds represent both a departure from tradition in origami and an area of active research. (Courtesy of the artists, Erik Demaine and Martin Demaine.)*

Origami: Unfolding the Future

Dana Mackenzie

ELEGANT IN ITS DESIGN and simple in its materials, origami (or paper-folding) has for more than 1400 years been a quintessential Japanese art. Generations of children and adults have learned how to fold classical figures like the crane, the jumping frog, and the flapping bird.

In the last few years, however, origami has taken off in a new, high-tech direction. Engineers are now designing retinal implants that will unfold inside the human eye, and enable people who have lost their vision to see again; robots that fold themselves out of a flat plane into three dimensions and then walk away; and "metamaterials" that change their physical properties on demand. Origami engineering uses new materials, breaks size barriers (creating objects that are scarcely larger than a mote of dust), and places a new emphasis on functionality rather than aesthetics. It also poses some new mathematical challenges.

Much of the explosion in research can be attributed to a visionary program of the National Science Foundation, called Origami Design for Integration of Self-assembling Systems for Engineering Innovation, or ODISSEI. Conceived by Glaucio Paulino of NSF's Engineering Directorate, ODISSEI funded eight projects in 2012 and five more in 2013, each one to the tune of $2 million. (The Air Force's Office of Scientific Research also kicked in some of the money.) "The ODISSEI program has made a huge difference," says Larry Howell of Brigham Young University, principal investigator on one of the grants. "As an engineer, when you do something as bold and audacious as origami, people ask you what you are talking about. You show them the funding source and it gives you instant credibility."

Some History of Mathematical Origami

Even before ODISSEI, origami science had been quietly gathering momentum for more than two decades. One of its pioneers in the United States was Robert Lang, a former engineer at the Jet Propulsion Laboratory and lifelong origami artist. In the early 1990s, Lang developed the first computer program for designing origami models, called TreeMaker. If you wanted to make an origami model for anything—say a beetle or a rhinoceros—and you could draw a stick figure of it, TreeMaker could calculate the folds required to bring your figure to life. (See Figure 1, next page.)

TreeMaker works by computing a circle packing of a square sheet of paper, which allocates each disk to make one appendage of the future model. The sizes of the disks correspond roughly to the sizes of the appendages, and adjacent disks correspond to adjacent appendages. Ironically, Lang says that

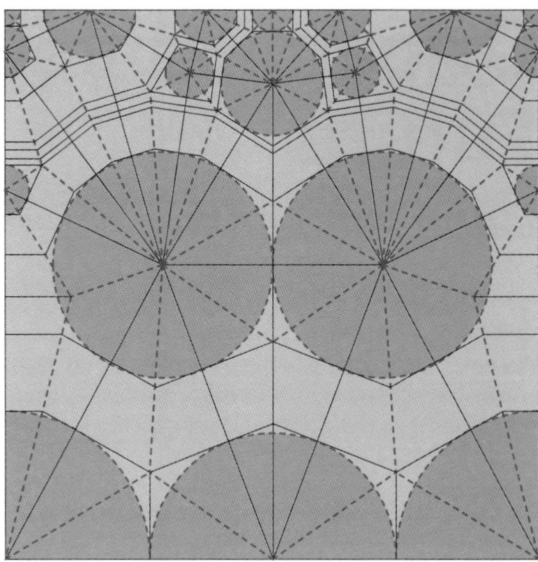

Figure 1. *"Roosevelt Elk, opus 358." It is believed that Robert Lang's "TreeMaker" software can print out a fold pattern to create an origami version of any 3-dimensional object, provided that the object can be drawn as a stick figure. However, this universality property has not been formally proved. (Left) An elk. (Right) The fold pattern used to create it. (Courtesy of Robert J. Lang, http://langorigami.com.)*

he seldom uses TreeMaker in his own art. "TreeMaker is a single idea, and I like to combine multiple concepts," he explains.

In 1998, Erik Demaine of the Massachusetts Institute of Technology (together with his father, Martin Demaine, and Joseph Mitchell of Stony Brook University) proved the foldability of *any* polygonal silhouette and *any* three-dimensional polyhedron from a single square of paper. In other words, if you can visualize it, you can fold it. However, the piece of paper may need to be enormous. As Demaine says, it would be interesting to know an efficient universal folding algorithm. The closest thing to date is another algorithm called Origamizer, written by Tomohiro Tachi of the University of Tokyo in 2008, for which Tachi and Demaine are close to proving universality.

Japan, the birthplace of origami, began to awaken to the potential of mathematical origami around the same time as people like Lang and Demaine in the West. Jun Maekawa, a physicist and software engineer, pioneered the study of crease patterns (the network of "mountain folds" and "valley folds" left in the paper when it has been folded into a model and unfolded again).

One of the most frequently cited theorems of origami mathematics is named after Maekawa. In classical origami, the final goal is usually an animal, such as a rhinoceros or a beetle, or an everyday object in three dimensions. But in many applications of origami, one needs to fold an initially large amount of material into a very small space. If the material could be folded flat, its volume would be essentially zero. (In the real world, of course, materials have thickness and volumes are not zero—but at least this is a target the designer can aim for.) So the question is: Which crease patterns can fold up into a flat, two-dimensional figure? Maekawa's theorem gives a necessary condition: at any flat-foldable vertex, the number of mountain folds and the number of valley folds must differ by two.

Figure 2.*Robert Lang with Miura-ori fold. (Top) Semi-folded Miura-ori. (Center) Miura-ori in compressed state. (Bottom) Miura-ori fully extended. The transition from folded to unfolded is accomplished by pulling the sides out in one smooth motion. (Photo courtesy of Dana Mackenzie.)*

Koryo Miura of the Japanese Institute of Space and Astronautical Science (ISAS) made another hugely important discovery about flat folding. As early as the 1970s, Miura began thinking about the problem of how a thin, flexible plate will buckle if you apply uniform compression to the outside. He discovered a prototypical solution, a sort of herringbone pattern, now called the Miura-ori or "Miura fold." The unfolded crease pattern for the Miura-ori looks like a pattern of zigzagging parallelograms. In accordance with Maekawa's theorem, there are three mountain

Itai Cohen. *(Photo courtesy of Cornell University Photography.)*

folds and one valley fold at every vertex, or vice versa. While tricky to fold from scratch, the Miura-ori is incredibly easy to fold when the paper is pre-scored. All you have to do is grab two corners and pull out; to fold it back in again, you take two corners and push in. (See Figure 2, previous page.)

These motions will cause the entire pattern, no matter how large, to expand or collapse at once. The Miura-ori is the ultimate solution to the map-folding problem. When you try to fold a conventional map, it nearly always folds up wrong. There are too many folds, and inevitably some of them will be done in the wrong order or in the wrong direction. With a Miura-ori map, on the other hand, folding or unfolding is foolproof, and it only takes a second.

It is often said that Miura developed his fold for the Japanese space program. That is not quite accurate, because his work started as the solution to a theoretical math problem. However, a solar array folded in the Miura-ori pattern flew aboard ISAS' Space Flyer Unit, launched in 1995, which demonstrated for the first time the practicality of using origami to stow a large array of solar panels inside a small spaceship.

Pop Goes the Defect

The Miura-ori is fundamental for anyone who wants to understand how movable origami works. Itai Cohen, a physicist at Cornell University, calls it "the hydrogen atom of origami," because it is one of the simplest possible tessellated folding patterns, yet still exhibits a variety of interesting and unexpected behaviors.

One remarkable feature of the Miura-ori crease pattern is its near-rigidity: it has one degree of freedom. A rigid object has no degrees of freedom. The one degree of freedom of the Miura-ori is provided by pulling out or pushing in on the corners. Once the motion is initiated in one place, the whole object has no choice but to follow. It has no unwanted, or "parasitic" motions.

The Miura-ori also has a remarkable physical property called a *negative Poisson ratio*. (Materials with this property are sometimes called *auxetic*, which is a little bit easier to say.) If you stretch most materials in one direction, they will tend to shrink in the transverse direction(s). The ratio between the amount of stretching in one direction and the amount of shrinking in the other(s) is called the Poisson ratio, and it is normally between 0 (e.g., cork) and 1/2 (e.g., rubber).

However, when you stretch a material with a negative Poisson ratio in one direction, it will stretch in the transverse direction(s) as well. The Miura-ori behaves exactly like this: pull the sides of the map, and the top and bottom will move out at the same time.

The physicist who is widely credited with inventing the first synthetic auxetic material was Rod Lakes of the University of Wisconsin. He described his material, a polymer foam with non-convex unit cells, in an article in *Science* in 1987. By now, many auxetic materials have been created. They have found application in sports wear, for instance, because they are good shock absorbers.

Jesse Silverberg. *(Photo courtesy of Jesse L. Silverberg.)*

In 2010, Jesse Silverberg, a lifetime origami enthusiast and graduate student in physics at Cornell, happened to go out to dinner after a talk with Chris Santangelo, a professor, and Marcelo Diaz, a fellow graduate student. The talk was about the geometry of folding so, Silverberg says, "I started folding something I had learned years and years ago"—namely, the Miura-ori. "I passed it to Marcelo and he passed it to Chris, and each of them said, holy crap!" What they had in front of them was an auxetic material—only it wasn't made of polymer foam, it was made out of simple paper. And it had been known long before 1987. They could be forgiven their holy-crap moment; it was like meeting someone from a different planet, and finding out that he already knows about the most sophisticated discoveries on your own planet.

Technically, the Miura-ori folded paper is not a material but a metamaterial: a material whose physical properties are partly based on its shape or configuration. An even simpler example of a metamaterial is reinforced cardboard. Three pieces of paper, stacked on top of each other, are not very stiff. But if you make the inside layer corrugated, the three layers together become stiffer and you have something that you can make boxes out of.

When the ODISSEI program was announced, Santangelo and Cohen sent in a proposal to study origami-based metamaterials, and they were funded in 2012. As it turned out, the surprises had just begun.

One of their collaborators, mathematician Thomas Hull of Western New England University, was particularly interested in misfolded Miura-ori (see Figure 3, next page). This may seem like a curious preoccupation, given the fact that the whole advantage of Miura-ori is the ease of folding it correctly. Nevertheless, Hull had noticed that if you deliberately convert some of the mountain folds in the Miura-ori to valleys and vice versa, you can sometimes get a flat-foldable pattern. He set his student Jessica Ginepro to work on counting the number of misfolded patterns in a 2-by-n Miura-ori, a 3-by-n, and so on.

Figure 3. *Misfolded Miura-ori. The orange vertex has been "popped" downward, creating a defect. (Figure courtesy of Jesse L. Silverberg, Arthur A. Evans, Lauren McLeod, Ryan C. Hayward, Thomas Hull, Christian D. Santangelo, and Itai Cohen.)*

Meanwhile, Santangelo, Cohen and Silverberg were looking at another way of mutilating a Miura-ori, by "popping through" its corners. If you push on one vertex of a Miura-ori, you can get it to pop through, converting a peak to a sink or vice versa. The popped Miura-ori is no longer flat foldable; the faces next to the popped vertex have to be bent. (The mere fact that it "pops" indicates that it is not a rigid motion. A rigid door swinging on a hinge does not pop.)

The first discovery that they made was that a Miura-ori with popped vertices is much stiffer than a non-popped one (see Figure 4). Maekawa's theorem mathematically forbids you to fold it flat, because the popping changes the number of mountain folds and valley folds at each vertex. Physically, if you try to fold it flat, its resistance to compression will skyrocket until one of the creases tears. "With paper, you can get it 10 to 100 times stiffer by introducing defects," Silverberg says. "Ten is what we were able to easily achieve, and around 100 we run into a limit because of the strength of the material. If we worked with plastic sheets, or something with a higher resistance to tearing, then 100 could be just the beginning."

The variable-stiffness property of popped Miura-ori makes it a tunable metamaterial. Unlike corrugated cardboard, whose stiffness is set once and for all, you can adjust the stiffness of

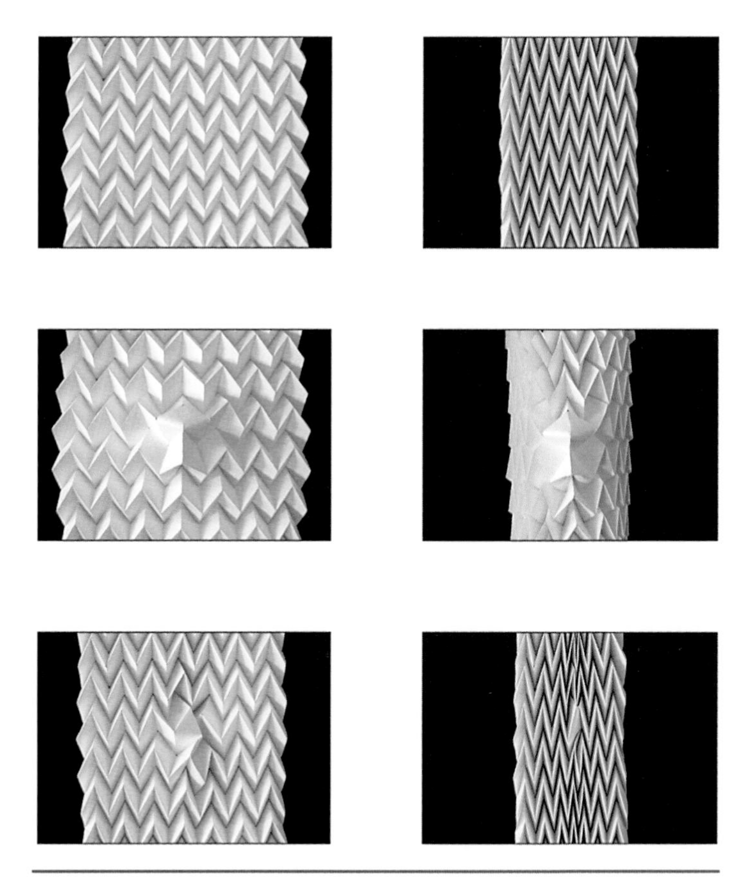

Figure 4. *(Top) Properly folded Miura-ori collapses easily. (Center) Defect in a misfolded Miura-ori resists collapsing. (Bottom) Pairs of defects can sometimes "cancel out" and create a flat-foldable design. (Figure courtesy of Jesse L. Silverberg, Arthur A. Evans, Lauren McLeod, Ryan C. Hayward, Thomas Hull, Christian D. Santangelo, and Itai Cohen.)*

a Miura-ori by increasing or decreasing the density of defects. You can also arrange the defects to make a hinge, or vice versa, to make a strut that is exceptionally resistant to bending. "Our idea is to use self-folding techniques to create robot limbs that change their mechanical properties," says Cohen. "You can fold a claw around something, make the claw rigid, and then pick it up."

Perhaps the most surprising discovery was the fact that defects can cancel each other out and make the Miura-ori flat-foldable again. The team realized that some of Hull's misfolded Miura-ori had a crease pattern that looked like two adjacent popped-through vertices. "Our spidey sense was tingling," Silverberg says. In fact, the equal-and-opposite pops restore the numbers of valley and mountain folds so that they once again obey Maekawa's theorem. Once the mathematical obstacle to flat foldability disappears, the increased stiffness also vanishes. "The idea of defects interacting came out of nowhere," says Cohen. "It was a total surprise. I had some initial intuition about the defects, but it wasn't until Jesse started working with the models that we discovered this. It was not what I predicted—it was much cooler."

Sam Felton. *(Photo courtesy of Peter York.)*

Self-Assembling Robots

The Cornell group reported their results on tunable metamaterials in *Science* magazine in August 2014. Not so coincidentally, their article appeared back-to-back with an article by another ODISSEI-funded group, co-organized by Demaine, Daniela Rus of MIT and Robert Wood of Harvard University. This paper, written by graduate student Sam Felton, reported on the first self-folding and self-activating robot (see Figure 5).

The Harvard/MIT group in fact progressed toward the robot in several stages. In earlier steps, they built a lamp that assembled itself but couldn't turn itself on, and an inchworm that could move on its own but required some human help in assembly. At the same time they were experimenting with several different shape-memory materials. According to Felton, their "secret sauce" is something that can be bought in a toy store—a heat-activated polymer called prestressed polystyrene or Shrinky Dinks. When heated to about 100 degrees Celsius, the Shrinky Dinks contract.

If a line is cut in a piece of cardboard and a piece of polystyrene is taped over it, the polystyrene acts as a hinge. As it contracts, it pulls the two sides of cardboard toward one another, creating a fold. The change is permanent, because the Shrinky Dinks cannot expand again after they have contracted. The angle of the fold can be controlled by the thickness and spacing of the two pieces of cardboard. Somewhat ironically for a project inspired by origami, the cardboard itself does not bend; it mostly adds stiffness and stability to the robot. The self-folding robot was made from five layers of material. On the outside were two layers of Shrinky Dinks (to allow both mountain folds and valley folds). Just inside them were two layers of cardboard, cut along each of the future folds. Finally, in the center of the sandwich was a single thin film that contains all the electronic wiring.

Felton first laid out the five-layer sandwich, with all the layers appropriately pre-scored and wired, and connected it to batteries and a microcontroller on top. This part of the assembly was done by hand, and took two hours. Then he turned the switch on. The microcontroller would execute a series of commands to heat up each wire in turn, activating the folds and causing the whole robot to rise up from the table like a salamander from the primordial ooze. This self-assembly phase took less than five minutes. After that, another series of pre-programmed commands caused the newly formed legs to move, and the robot trundled away under its own power. At a clip of 3 meters a minute, or about an eighth of a mile per hour, it couldn't win the Olympics but it could win a race with a snail.

Of course the self-folding robot was a proof of principle rather than a practical device. However, it has potential for application in two realms that pose challenges for traditional methods of assembly. One would be space missions, where it would be advantageous to store a rover as a flat panel in flight, then unfold and deploy it after arrival. With no humans on board, the rover would have to be self-assembling and self-activating.

A second application would be microscopic devices. For such devices, conventional assembly methods don't work because, as Cohen says, "We don't have good nano-screws yet." Three-dimensional printing, though it is all the rage in techie circles,

Figure 5. *Self-folding robot developed by Sam Felton, collaborating with Erik Demaine, Daniela Rus, and Robert Wood. (Figure courtesy of Wyss Institute at Harvard University.)*

still doesn't work very well for small devices. But the tools for two-dimensional microfabrication have been developed to a very high degree of perfection, thanks to the computer industry. Thus the idea of origami engineering is to make things in two dimensions, which we are good at, and then fold them up into the third dimension. "Folding is a lot cheaper and faster than 3-d printing," Felton says.

Howell's group at Brigham Young University, also funded by an ODISSEI grant, has demonstrated origami microfabrication for a practical problem: how to inject DNA into mouse eggs for gene therapy. They developed a six-bar mechanism that can be etched onto a silicon wafer, then folded up into three dimensions by a distance of exactly 45 microns—half the diameter of a mouse egg. Thanks to perfect control over the height of the needle and the depth of penetration, they performed the injec-

....[T]he tools for two-dimensional microfabrication have been developed to a very high degree of perfection, thanks to the computer industry. Thus the idea of origami engineering is to make things in two dimensions, which we are good at, and then fold them up into the third dimension.

tions more efficiently than is possible by hand. The survival rate of the injected embryos (71 percent) was almost as good as for non-injected embryos (80 percent).

From Origami to Viruses

Nature, too, may be an origamist at heart. Several organs in the body, such as the brain and the intestine, are made up of intricately folded two-dimensional tissue. "Remarkably, such folded geometries self-assemble with remarkable precision," says David Gracias, a biomolecular engineer at Johns Hopkins University. Gracias' group is trying to understand this folding self-assembly process at all scales, from molecules like proteins to organs.

Together with Govind Menon, an applied mathematician at Brown University, Gracias uses an experimental system in which synthetic origami cutouts self-assemble into polyhedra. Gracias developed the system as a post-doctoral fellow in the lab of George Whitesides at Harvard University. The model structures start as interconnected polygonal nets, which upon heating spontaneously fold themselves up into three-dimensional polyhedra (such as cubes, octahedra, and dodecahedra). The nets are made from metal panels joined by hinges made from a low-melting-point solid, and they have been made as small as a few hundred nanometers, roughly the same size as many viruses. When immersed in a sufficiently hot solvent the solid hinges melt and pull the panels together by surface tension. The self-folding process is, however, not foolproof. After all, no wires or active control are involved, and the heating is done by a uniform bath. Typically only about 30 percent of the dodecahedra are defect-free, and Gracias' team has yet to produce a defect-free icosahedron. However, the fact that it works at all—the idea that you can just throw a few seeds into the solvent, turn on the heat and watch them spontaneously fold up into beautiful Platonic solids—is amazing. "It took them ten years of work to get to the state where folding simple shapes could be done by an undergraduate," Menon said.

That was the point when Gracias came to Menon, who had been a friend of his since the two were in college. Gracias knew that many different nets can fold up into the same polyhedron. There are 11 possible nets for the cube, 43380 for the dodecahedron, and an estimated 2.3 million for a truncated octahedron (which is slightly less symmetric, an Archimedean rather than Platonic solid). Some of these nets give a higher yield of defect-free polyhedra than others, and Gracias wanted to know why. "It's the sort of question that emerges when a technology is moving away from proof of concept to repetitive manufacturing," Menon says. "That's a good stage for a mathematician to get involved."

Menon was not sure what he could offer at first. "My first reaction was that there is no math here," he says. The cube was a little bit too easy to be interesting, because it could be solved by going through all 11 cases. But the 43 thousand cases for the dodecahedron were more interesting, and Menon's students discovered that simple, quantifiable characteristics of the precursor net could be used to predict how well it would fold.

"That's when I began to take this more seriously," Menon said. "Why does this work? What exactly is the process of folding, and how do I model it?"

Menon realized that he needed to look more carefully at the "assembly pathways"—the partly folded polyhedra that the models pass through on the way to being completely folded (see Figure 6). These pathways form a network structure called a graph, and Menon discovered that the more successful nets take pathways through the network that pass through only a handful of the possible intermediate steps. These same pathways had been seen in Gracias' experiments with the synthetic polyhedra. The "good" intermediates are the ones that are more rigid, and they could be predicted by computing the degrees of freedom. In particular, the second-to-last step always involves the formation of two rigid hemispherical caps, which then click together to form the (roughly) spherical model.

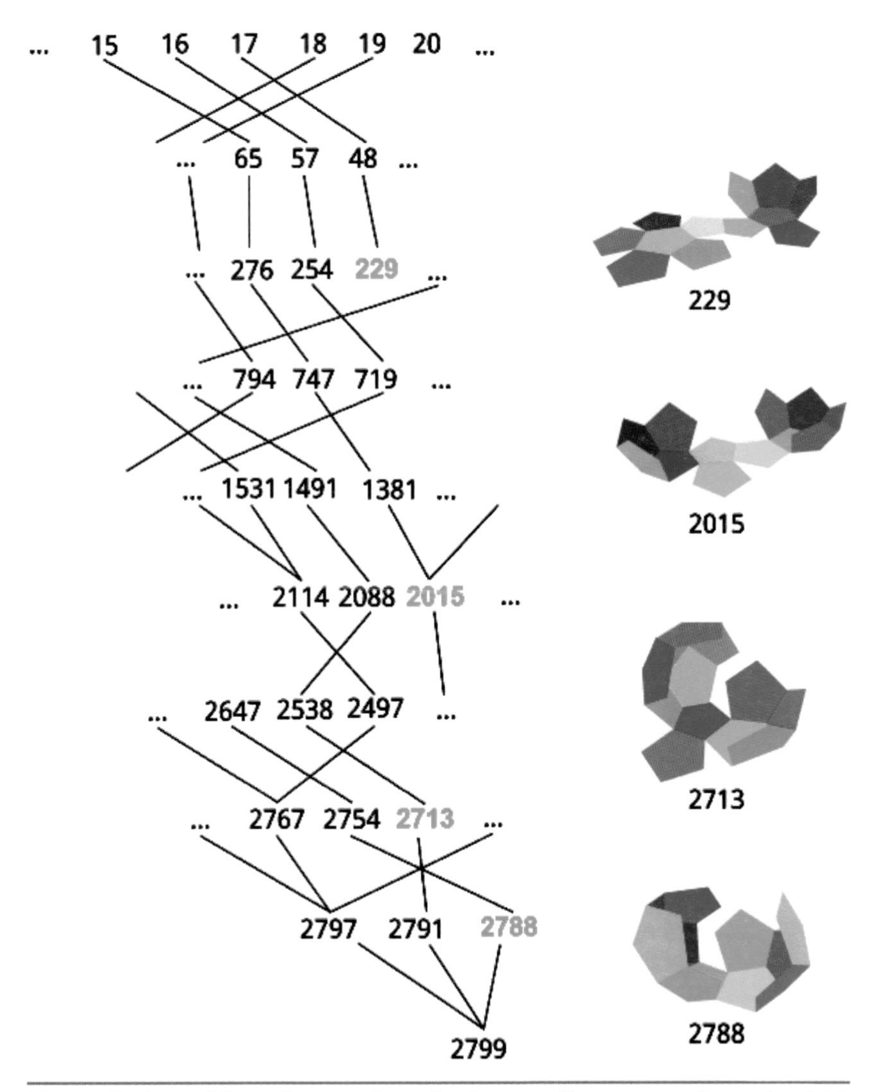

Figure 6. *Assembly pathways show the intermediate stages through which a polygonal net passes as it self-folds into a doecahedron. (Figure courtesy of Govind Menon. Image credit: Ryan Kaplan.)*

This has been a central problem in biology for more than half a century: How do biological systems, from organisms down to proteins, assemble themselves into the correct form much more quickly and reliably than physical models suggest they should?

What do self-folding polyhedra have in common with viruses? Since the 1950s, biologists have known that viruses have an outer shell, or capsid, that is shaped like a polyhedron. One in particular, called MS2, has icosahedral symmetry (the same kind of symmetry as the models that Menon and Gracias are experimenting with). The real viruses probably assemble themselves through an attachment process rather than a folding process. For this reason, the assembly pathways may differ somewhat from the pathways for self-folding polyhedra. Nevertheless, Menon believes that the same general principles will hold true. That is, the virus will choose from a relatively small number of preferred pathways, possibly the ones that generate the most rigid intermediate steps. If you can understand the preferred pathways, you can understand how the virus assembles itself with no defects.

This has been a central problem in biology for more than half a century: How do biological systems, from organisms down to proteins, assemble themselves into the correct form much more quickly and reliably than physical models suggest they should? "The central idea is assembly pathways," Menon says. "You have a configuration space, and pathways going between different states, and you want to understand, out of the set of all possible pathways, which are the most important. This is something that I wish more mathematicians knew about."

Artificial Vision

Often it has been said that beauty is in the eye of the beholder. But never has it been so literally true as in the experimental project to develop an artificial retina.

Together with Azita Emami and Sergio Pellegrino, who are both engineers at Caltech, Robert Lang has been working on a way to implant an array of electrodes into the eyes of people who have lost most of their vision due to macular degeneration. The electrode technology was developed by Mark Humayun, an ophthalmologist at the University of Southern California, who implanted flat arrays of 60 electrodes into patients' eyes. In these patients the cells in the retina have died, but the nerve fibers behind them are still alive and well. Humayun's idea was to use a camera mounted to the patient's eyeglasses to record a scene, and then transmit the picture to the artificial retina. The electrodes would pass the signals along to the patient's optic nerves and eventually to the brain. The treatment worked—the patients reported seeing an array of spots. However, a 60-pixel array of colored spots falls far short of the beautiful and complex world around us. So in 2012, Emami took up the challenge of developing a larger, 512-pixel array.

Two problems immediately presented themselves. First, how could the larger array of electrodes be introduced into the patient's eye? The original 60-electrode chip already required a large incision. She and Humayun realized that a better way would be to put the array on a flexible material and fold it up so that it could pass through a much smaller incision. Once inside the eye, the array could be unfolded to its full size.

The second problem is less obvious. A larger grid would have to fit more snugly against the back of the eye than the flat chip that Humayun had used. Otherwise the electrodes would not be able to send their signals to the appropriate nerves. "They need to set it within 20 microns of the back of the retina," Lang says. But the back of the eye is curved, not flat. And if you've ever tried to wrap a doorknob in wrapping paper, you realize that it is exceedingly difficult to form a curved surface out of flat paper, or any other flat material.

That's when they turned to Lang—an engineer and origamist extraordinaire. "I've got an algorithm for that," Lang told them.

It's true that there is a fundamental mathematical limitation to what paper can do, called the Gaussian curvature. Back in the nineteenth century, the versatile German mathematician Karl Friedrich Gauss proved what he called his "theorema egregium" ("remarkable theorem"), which says that a non-stretchy surface can never change its Gaussian curvature. Paper is extremely non-stretchy. If you try to stretch it, it tears. Paper also starts with a Gaussian curvature of zero, because it lies flat in a plane. On the other hand, an eyeball, or any spherical surface, has positive curvature. So in principle, you shouldn't be able to wrap the outside of a doorknob or the inside of an eye with paper, or any other non-stretchable flat material.

But Lang knew that there are all sorts of asterisks attached to this theoretical impossibility. First, paper *can* bend in *one* direction—it just can't curve in two orthogonal directions at the same time, the way the surface of a sphere does. By putting together enough bending panels, you can make a pretty good approximation to a sphere—for example, a Chinese lantern.

Second, Lang knew about the amazing accomplishments of his fellow origami artists in making objects with curved folds. Most people don't realize that you can fold a piece of paper along a curve. But as early as the 1920s, Josef Albers, an artist with the Bauhaus movement, started experimenting with curved folds. Ron Resch, a computer scientist, and David Huffman, a mathematician well known for developing "Huffman codes," also made some beautiful, gently rounded sculptures using methods that are still not completely understood. In the 1990s, Chris Palmer discovered you can get something very close to a sphere by folding a piece of paper with curved pleats (see Figure 7, next page). A decade later, Jeanine Mosley discovered another way with smaller, less visible pleats. Lang had studied the mathematics behind their work, and realized that the same pleating techniques could be applied to the artificial retina project.

Lang emphasizes that the project still has a long way to go. Of course the retinas will be made from a suitable biocompatible material, not paper. But paper is a wonderful material—there is nothing else quite like it. It may not be easy to execute folds like Palmer's or Mosley's with a different material, inside a patient's eye. "The origami model is only 5 percent of the work, and engineering is 95 percent," Lang says. However, he adds, what the origami gives to the engineer is a starting point. "You need a structure to help you focus your investigation."

Most people don't realize that you can fold a piece of paper along a curve.

Figure 7. *Robert Lang's version of Chris Palmer's folded sphere. Pleated origami with curved folds can achieve remarkably good approximations of surfaces that are in theory not foldable with a flat piece of paper. (Figure courtesy of Dana Mackenzie.)*

A Look Ahead

While the ODISSEI grants have tended to put applications of origami into the foreground, the ancient art continues to be a fertile source of intriguing problems in pure math. For instance, curved folds and curved surfaces are a very hot topic. Demaine's foldability theorems settle what can be done (in theory) within the universe of flat folds and planar faces, but a similar understanding of curved folds remains far away. Curved folds set up stresses within the paper that turn it from a passive to an active participant in creating a shape. Mathematicians like Demaine are still trying to figure out what is going on.

Other areas remain wide open for mathematical investigation. Lang mentions two of them. "First, we need a good mathematical treatment of thickness," he says. Nearly all the theorems proven to date assume paper with zero thickness—a convenient fiction, but a fiction nevertheless. As origami moves to smaller sizes and to different materials, it will have to deal with thick folds. Howell, for instance, used finite-element analysis to predict the behavior of his microinjector. But finite-element analysis, like any purely computational technique, has its limitations. It produces answers, but not necessarily understanding.

Second, Lang says, "I think there is still a lot to be done in the area of rigid foldability." Paper bends, but some other important materials don't. Engineers working with metal or glass may need to limit themselves to folds that can be accomplished without bending. Rigidity also seems to be an important consideration in Menon's assembly pathways, and the semi-rigidity (one degree of freedom) of the Miura-ori is likewise essential to its usefulness. At present there is still not a good theory to predict how many degrees of freedom a folded object will have.

Origami continues to surprise us with its ingenuity and creativity. It is leading to new mathematical theorems and, even better, to new problems. At the same time, artists continue to find ways to push its mathematical limits, and sometimes fool us into thinking they have done the impossible. Materials such as shape-memory polymers and alloys have opened the door to self-folding and to new engineering applications. Techniques such as laser etching make it possible to fold designs with a precision never before imaginable. Computer folding algorithms expand our ability to design new pieces of origami art. Folded metamaterials hint at a future that is still hard to grasp, when the very things around us will have a chameleon-like ability to change their physical characteristics. And all of these steps are leading us to a greater appreciation of how nature folds up a virus, a protein or an organ. For an art that is more than a millennium old, origami seems to be full of ideas for twenty-first century science.

Origami continues to surprise us with its ingenuity and creativity. It is leading to new mathematical theorems and, even better, to new problems. At the same time, artists continue to find ways to push its mathematical limits, and sometimes fool us into thinking they have done the impossible.

Buchstab function. *(Courtesy of Princeton University Press. From Andrew and Jennifer Granville,* Prime Suspect: A Mathematical Sciences Investigation, *Princeton University Press, forthcoming.) In reality, number theorists do not use sieves to solve crimes, as in this forthcoming graphic novel by number theorist Andrew Granville. But they do use them to find primes. Sieves, and the same Buchstab's function mentioned above, were instrumental in the proof that there exist infinitely many pairs of prime numbers in which the larger prime is at most 246 greater than the smaller prime.*

Prime Clusters and Gaps: Out-Experting the Experts

Dana Mackenzie

ONE OF THE MOST ENDURING (AND ENDEARING) myths in mathematics is that of the solitary genius. After years of working alone and unappreciated, the hero emerges with a brilliant new theorem. The mathematical world bows and lavishes him or her with honors. The names Ramunajan, Wiles and Perelman come to mind. The truth, however, is seldom so romantic. Many (if not most) breakthroughs emerge from collaborations. Even if the capstone theorem is the work of one individual, it will always build on the work of many. Finally, the person who makes the breakthrough almost never emerges from complete obscurity.

Recently, though, something close to the fairy-tale version of mathematics actually did come true. In 2013, Yitang (Tom) Zhang, an émigré from China and mathematics lecturer at the University of New Hampshire, made the most significant advance in decades toward solving an unsolved problem in number theory called the twin prime conjecture. Overnight he became a celebrity, profiled in publications like the *New Yorker*. In 2014, the MacArthur Foundation awarded him a well-deserved "genius grant."

The man behind the commotion turned out to be one of the most soft-spoken and humble celebrities in history. But his passion for mathematics was genuine, and his perseverance in the face of adversity could be an example to anyone, especially those who for reasons of birth or fortune have not landed in the best places for a mathematical career.

Born in Shanghai in 1955 to a college teacher, Zhang was forced to live for some time in the countryside during the Cultural Revolution. However, by the late 1970s higher learning was once again respectable in China. Zhang attended Peking University and graduated as one of the most gifted math students in his class. Like many other top Chinese students at the time, he came to America to study for a Ph.D., which he earned from Purdue in 1991.

That was when Zhang caught a bad break. The early and mid '90s were a unique period in American mathematics, a time when even the best students struggled mightily to find a job. The reasons are not entirely clear. Some attributed it to the huge influx of talented mathematicians from the former Soviet bloc, which had recently collapsed. Others attributed it to retirements in the professoriate that were not replaced. From 1992 to 1996 the unemployment rate among new Ph.D. mathematicians was over 10 percent. For probably the only time in history, it was better not to have a high-school diploma than to have a Ph.D. in mathematics.

Yitang (Tom) Zhang. *(Photo courtesy of George Csicsery. ©2014 All rights reserved.)*

Zhang came to this dreadful job market with a big strike against him: through no fault of his own, his thesis advisor would not write a letter of recommendation for him. The road to a career in mathematics seemed to be closed. But Zhang didn't give up. He intermittently worked in a Subway restaurant, owned by a friend, during the 1990s. In 1999 he finally obtained a job at New Hampshire as a lecturer, a position that typically offers year-to-year contracts, heavy teaching loads, and little prestige. He was still working in that capacity up to April of 2013. That was when "Professor Tom" submitted a paper for publication that, a month later, suddenly made him the most famous mathematician in the world.

It was a real-life rags-to-riches story, and mathematicians were quick to acknowledge how remarkable a feat it was. "Zhang did it by out-experting the experts," said Andrew Granville, a number theorist at the University of Montreal. Granville compares him to Albert Einstein, who developed the special theory of relativity while working in a patent office.

Amazing as it was, Zhang's story was not the whole story. His announcement set off a frenetic few months in which mathematicians bettered his result literally day after day (mostly in very minor ways). Then, in October, another mathematician came out of the woodwork. James Maynard, a post-doc who had just received his doctoral degree from Oxford, out-Zhanged

Zhang with a proof that was both simpler and more general. Best of all, it was *different* from Zhang's, which meant that it was possible to combine the two to get even better results. By the time the dust settled in mid-2014, number theorists were vastly closer to one of their most cherished goals, understanding the distribution of gaps between prime numbers. Although Maynard did not get quite the worldwide publicity that Zhang did, number theorists knew that he too deserved a large share of the credit.

The Distribution of Prime Numbers

Prime numbers are numbers like 2, 3, and 17, which are divisible only by themselves and 1. They are the building blocks of the number system, because every other positive number (except 1) can be separated in a unique way into a product of primes.

In many ways, prime numbers look as if they are distributed randomly among the other integers. They are certainly not distributed evenly. There are many clumps of several primes close together (say, 101, 103, 107, and 109) and there are also barren patches with no primes (for example, the gap between 113 and 127). Gaps between consecutive primes satisfy certain rather obvious rules. A prime gap can never have length 1, with the sole exception of the gap between 2 and 3. That is because any consecutive pair of numbers must include an even number, and this number (unless it equals 2) is not prime.

James Maynard. *(Photo courtesy of George Csicsery. ©2014 All rights reserved.)*

Zhang did not solve the Twin Prime Conjecture. That is, he did not show that 2 is a de Polignac number. But what he did was nearly as important. He showed for the first time that *de Polignac numbers exist.*

Likewise, it is impossible to find a cluster of three primes that are consecutive odd numbers, except for 3, 5, and 7. The argument is essentially the same: one of the three in such a cluster would have to be divisible by 3.

Number theorists believe that, aside from these "obvious" restrictions based on divisibility arguments, there are no other constraints on prime gaps or prime clusters. That is, primes that are two units apart, called "twin primes," never end. This statement, which has never been proved, is called the Twin Prime Conjecture. (The largest *known* twin primes at this writing are the 200-thousand-digit numbers, $3756801695685 \times 2^{666669} \pm 1$, but no one believes they are the last.) Similarly, "admissible triples" of primes, such as p, $p + 2$, and $p + 6$, keep happening forever, and there are similar versions of the conjecture for quadruples, quintuples, etc.

It's unclear who first thought of the Twin Prime Conjecture; the first published statement of it was by the French mathematician, Alphonse de Polignac, in 1849. It may have been around as "folk wisdom" even before then. De Polignac went farther and conjectured that there is nothing special about the number 2: *every* even number appears as a prime gap infinitely often. Granville has proposed calling such numbers (i.e., numbers that appear infinitely often as prime gaps) *de Polignac numbers.*

Zhang did not solve the Twin Prime Conjecture. That is, he did not show that 2 is a de Polignac number. But what he did was nearly as important. He showed for the first time that *de Polignac numbers exist.* In fact, the smallest one is less than 70 million. "He could have gotten it smaller, but the important thing is that there exists such a number. That's a huge step," said John Friedlander, a number theorist at the University of Toronto.

Other mathematicians chipped away at Zhang's upper bound throughout the summer of 2013. Their *de facto* leader was Terry Tao, a mathematician from UCLA who was already well known for his work on prime numbers in arithmetic progressions. "Initially I didn't want to get involved," said Tao, "but I soon couldn't resist the game of working for a few minutes to see if I could knock a percentage point or two off the bound." It was a situation that rarely occurs in mathematics, like a vein of ore so rich that you could get world-class nuggets just by reaching in and grabbing.

Together with Scott Morrison, Tao organized an informal crowd-sourcing group called Polymath 8 to keep track of all the improvements in Zhang's argument. (There had been seven previous Polymath, or "massively collaborative mathematics" projects, in various areas of mathematics starting in 2009.) Between May and October, the Polymath collective brought Zhang's estimate for the first de Polignac number down by leaps and bounds, first into the millions, then the thousands, and finally by late October, to 4680.

All of these advances were like minor aftershocks to the Zhang earthquake. But in October, a completely independent earthquake happened. At a conference in Oberwolfach, Germany, Maynard announced that he could prove the existence of a de Polignac number less than or equal to 600. It wasn't just the size of the improvement that was remarkable. It was the fact that he had a completely different method. He had resuscitated an old technique that had been thought not to work, called a "multidimensional sieve." His result was also much stronger than Zhang's. Zhang's proof only shows the existence of de Polignac numbers for *pairs* of primes. But Maynard's proof also works for triples or for larger clusters. For any integer k, you can find infinitely many bouquets of k primes that can each fit inside your hand (assuming you have a large enough hand).

Like Zhang, Maynard was not quite in the circle of top number theorists. His advisor, Roger Heath-Brown of Oxford, certainly belonged to the inner circle, but Maynard himself was scarcely out of graduate school. As Granville says, he was too young to know that what he was doing was not supposed to be possible—or at least too young to let it stop him.

Also like Zhang, Maynard worked mostly by himself, and separately from the Polymath project. In fact, he had started working on prime gaps *before* Zhang's electrifying announcement. Though Polymath had come up with incremental improvements, both of the year's major breakthroughs came about through old-fashioned, nineteenth-century-style solitary genius.[1]

Not that Polymath 8 was a failure! It was anything but. The true value of the collaborative turned out to lie not in making new discoveries but in assimilating and disseminating them. Almost immediately after Maynard's announcement a Polymath 8b sprouted, with the idea of combining Maynard's and Zhang's techniques. Maynard actively participated in it; Zhang did not, because he had moved on to other research problems. At the time of the writing of this chapter, the first de Polignac number is now known to be equal to 246 or less. The Twin Prime Conjecture remains unproven and appears out of reach—but certainly not as far out of reach as it once did.

The true value of the collaborative turned out to lie not in making new discoveries but in assimilating and disseminating them.

One Million Shades of Gray

Both Zhang's work and Maynard's used the concept of *sieves*, and built on methods that had been developed about eight years earlier by Dan Goldston, Cem Yildirim, and János Pintz. A sieve is a method of finding primes in bulk, rather than one at a time, and as such it is especially good for problems that involve counting large numbers of primes.

The simplest version of a sieve is called the sieve of Eratosthenes (see Figure 1), discovered in ancient Greece. In the sieve of Eratosthenes, you first list all the integers after 1. Whenever you come to an integer that hasn't been crossed out yet (i.e., 2), you know that number is a prime. Then you cross out all of its multiples. The next number that hasn't been crossed out yet (this time, 3) is another prime, and you cross out all of its multiples. Repeat as many times as you want.

While the sieve of Eratosthenes is straightforward and 100 percent accurate, it is also rather clumsy. In the pre-computer era, if you wanted to know the number of primes less than 1 million, you would be computing for a long time. (There are exactly 78,498 of them.) By contrast, mathematicians know

[1]It should be noted that Tao also tried the multidimensional sieve idea and proved a result similar to Maynard's. With remarkable generosity, Tao agreed to let Maynard claim it first in print so long as he also acknowledged Tao's independent work.

very good approximate estimates that involve simple functions. The number of primes less than 1 million is roughly $1,000,000/\log(1,000,000) \approx 72,382$. An even better estimate is

$$\int_2^{1000000} 1/\log x \, dx \approx 78,628\,.$$

(Here "log" represents the natural logarithm function, not the logarithm to base 10.)

This approximate prime-counting method, called the Prime Number Theorem, has some very important consequences for the worldview of number theorists. It suggests taking a less black-and-white view of prime numbers. The Prime Number Theorem means that if you randomly pick a number somewhere around 1,000,000, there is about a 1-in-13 chance (or 1-in-log-1,000,000 chance) that it is prime. If you color primes black and non-primes white, and then blur your eyes, every number around 1 million looks light gray (or 1/13 black).

Modern sieves take this idea a step further. In a sieve invented by Atle Selberg, instead of erasing multiples of 2 completely, you just make them 1/2 lighter. When you come to a multiple of 3, you make it 1/3 lighter. If a number is divisible by 6, it has already been partially erased *twice*, so you compensate by making it 1/6 darker. Like Eratosthenes' original sieve, this procedure can be repeated through as many cycles as desired.

The darkness of a number after it has gone through Selberg's sieve is called its *weight*. In the black-and-white world of Eratosthenes' sieve, the only weights are 0 (not prime) and 1 (prime),

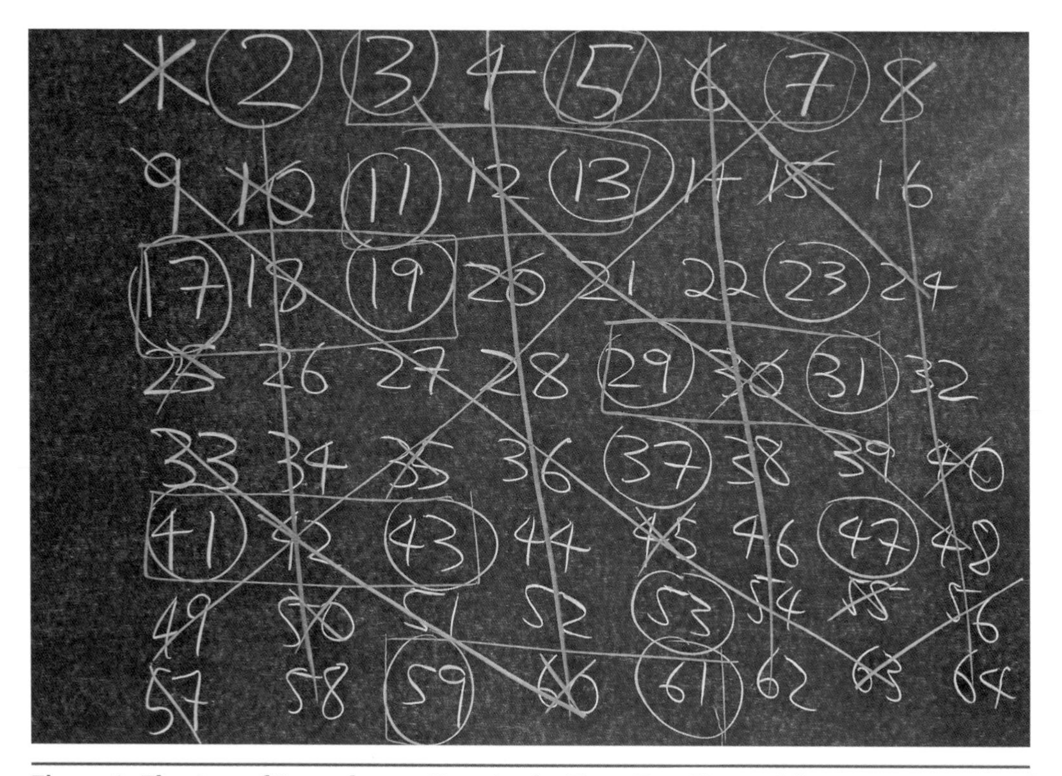

Figure 1. *The sieve of Eratosthenes. (Drawing by Terry Tao. Photo of illustration courtesy of George Csicsery.) The sieve of Eratosthenes is a method for finding prime numbers that was discovered by the ancient Greeks. More modern versions use fractional weights to identify numbers as being probably prime or probably not prime.*

but in the gray world of Selberg's sieve, they can be anywhere between 0 and 1. The virtue of using a sieve with weights is much like the virtue of blurring your eyes in the Prime Number Theorem. Number theorists can approximate the weighted number of primes (or prime pairs, as we shall see) using ordinary smooth functions, with a very high degree of accuracy.

In 2005, Goldston, Pintz, and Yildirim published a paper that turned the subject of prime gaps from a curiosity to a field of active research. They did this, first, by introducing a more flexible weighting method—a variant of Selberg's weights that tends to concentrate the weight onto numbers that have, say, 4 or fewer prime factors. Second, they applied these weights to products of "admissible tuples." Remember, for example, that n, $n+2$, and $n+6$ form an admissible triple of numbers—there are no divisibility restrictions keeping all three of them from being prime. What happens if you apply to their product, $n(n+2)(n+6)$, a weight function that concentrates on numbers with 4 or fewer prime factors? If the weight is close to 1, you can conclude that $n(n+2)(n+6)$ "probably" has at most 4 prime factors, which means that two of the numbers n, $n+2$, or $n+6$ are "probably" prime. Thus you "probably" have a prime pair that is at most 6 units apart.

In 2005, Goldston, Pintz, and Yildirim published a paper that turned the subject of prime gaps from a curiosity to a field of active research.

Of course, for any individual value of n, this "probable" conclusion doesn't do you much good. Either it is part of a prime pair or it isn't, with no probability about it. However, the real value of the probabilistic method is that you can average these weights over many numbers n, and you can estimate these averages as well as their error. Goldston, Yildirim, and Pintz developed a new way of estimating the averages by splitting the numbers into bins called "arithmetic progressions."

An arithmetic progression is a set of numbers that are separated from each other by a constant amount. For example, $(1, 7, 13, 19, \ldots)$ is an arithmetic progression with common difference 6, as is $(5, 11, 17, 23, \ldots)$. All primes except 2 and 3 fall into one of these two bins. One of the greatest theorems of nineteenth-century number theory, Dirichlet's theorem, says that over the long run, each bin will get roughly an equal share of primes.

But how long is "the long run"? That turned out to be the crucial question for the success of the "GPY method." In independent papers, both published in 1965, Enrico Bombieri and Askold Vinogradov showed that if the bins have common difference q (where q is less than $x^{1/2}$—in other words, quite a lot less than x), then each bin contains roughly an equal number of primes less than x. It is widely believed, however, that Bombieri and Vinogradov's result is not the last word on the subject. In 1968, Peter Elliott and Heini Halberstam conjectured that the Bombieri-Vinogradov theorem holds if $q < x^{\theta}$, for any number $\theta \leq 1$. The number θ is called the "level of distribution" of the prime numbers. A level of distribution of $1/2$ (the Bombieri-Vinogradov theorem) means that it takes primes a relatively long time to become equally distributed among the bins. An exponent close to 1 (the Elliott-Halberstam conjecture) would mean that they are equally distributed almost right off the bat.

The "GPY method" thus led to a best-case and worst-case scenario. In the worst-case scenario, where the level of distribution is $1/2$, the three mathematicians were unable to show

The GPY method thus pointed to a clear route forward: Prove the Elliott-Halberstam conjecture. Or at least prove that the level of distribution of primes is strictly greater than 1/2.

the existence of de Polignac numbers. However, they just barely failed. They did show that the *minimum gap* between any pair of primes greater than n increases more slowly than the *average* gap (which is roughly $\log n$). This was already a major improvement over any known result. Even better, they showed that any level of distribution strictly greater than 1/2 would put a fixed ceiling on the minimum gaps. In the best of all best-case scenarios, if the level of distribution is greater than 0.971, then the ceiling is 16. In other words, there is a de Polignac number less than or equal to 16. If true, this would be amazingly close to the Twin Prime Conjecture.

One of the most succinct summaries of the GPY theorem came in an article written by Jordan Ellenberg for *Slate*: "Random is random." That is, randomness (in the form of equal distribution of the primes among arithmetic progressions) begets more randomness (in the form of prime pairs). The last point may not seem quite obvious, but the idea is that primes *should* clump up sometimes, unless there is some hidden force or pattern (a non-random behavior) that pushes them apart.

The GPY method thus pointed to a clear route forward: Prove the Elliott-Halberstam conjecture. Or at least prove that the level of distribution of primes is strictly greater than 1/2. Unfortunately, from 2005 to 2013 nobody was able to do that, although quite a few top number theorists tried. The stage was set for Zhang's and Maynard's twin breakthroughs.

Completing the Argument

Zhang says that he started working on prime gaps after he read the paper by Goldston's team. He started where they had left off, by trying to improve on the level of distribution. Remember that this number is an estimate of how long it takes for the primes to apportion themselves evenly into arithmetic progressions with a common difference q. It took three years for Zhang to come up with the idea that unlocked the problem for him: the process of arriving at an equal distribution among the bins happens just a little bit faster if q is "smooth," i.e., it has no large prime factors. The primes behaved in this case as if they had level of distribution $1/2 + 1/584$. This ever-so-tiny improvement was, as Goldston's team had showed, enough to give a finite bound on prime gaps.

Zhang's work was vastly more difficult than it sounds from this description, calling on very deep theorems in number theory due to Pierre Deligne and André Weil. One difficulty is that Zhang was running fewer iterations of the sieve. As Goldston says, it was therefore less efficient but with better control over the errors, because using only smooth numbers makes the weighted sums in the GPY sieve behave more like a smooth function. Zhang had to show that on balance, the reduction in the error outweighed the information lost by throwing out some of the sieve iterations. In particular, he had to show that the non-smooth numbers were not too common. (See Figure, the **Buchstab function**, page 18, from a forthcoming graphic novel by Granville. The story is purely fiction, but the "Buchstab function" is real, and used to count non-smooth numbers.)

Maynard took a radically different approach, inspired by a *previous* paper of Goldston and Yildirim (without Pintz). Goldston and Yildirim had originally, in 2003, tried to use a multidimensional sieve. For example, if you are searching for prime triples, you can use separate weights for n, $n+2$, and $n+6$ respectively; the dimension of the sieve is the number of different weights. Unfortunately, their argument had a flaw in it. Pintz rescued their proof with a one-dimensional sieve, which computes a single weight for the product $n(n+2)(n+6)$, as explained above. This was the same GPY sieve that Zhang used. In a case of guilt by association, number theorists viewed the old multi-dimensional sieve with suspicion because it had backfired on Golston and Yildirim the first time.

But Maynard, with time on his hands after his thesis defense, returned to it and found a way to make it work for all k-tuples. Instead of using the relatively simple weight functions that GPY used, he undertook a systematic search for the *best* weight functions. He didn't find them. However, when $k = 105$ (or larger) he was able to find weight functions that were *close enough* to best that they worked where Goldston and Yildirim's had failed. The smallest admissible 105-tuple turned out to be $(n, n+10, n+12, n+24 \ldots, n+598, n+600)$. There are infinitely many choices of n such that this 105-tuple will contain a pair of primes, and those two primes will therefore create a prime gap less than or equal to 600. The current best result, as of this writing, is based on an admissible 50-tuple, which goes $(n, n+4, n+6, n+16, \ldots, n+244, n+246)$. By the same reasoning, it gives infinitely many prime pairs with a gap less than or equal to 246.

What is the limit of the GPY-based sieve methods? How small can the smallest de Polignac number be? "If you take a very restrictive view of what my method is, then 12 is the best you can get," says Maynard. For sieve methods in general, the number 6 is seen as a more fundamental limit, due to something called the "parity problem." Roughly speaking, sieves only detect about half as many primes as they should. Suppose, for instance, that the Prime Triples Conjecture is actually true. Then there would be infinitely many triples $(n, n+2, n+6)$ that consist of three primes. Sieves will detect, on average, half of these (in a best-case scenario), so they will detect an infinite sequence of triples that average 3/2 primes. But if the *average* is 3/2, there must be many triples that contain 2 detected primes. So the sieve will detect infinitely many prime *pairs* that are at most 6 units apart.

But the argument fails if you want to show that there is an infinite sequence of pairs that are two units apart. Even if such a sequence exists, sieve methods will (in a best-case scenario) detect half of them, so you will have an infinite sequence of pairs that contain an average of 1 detected prime each. If that is the case, then it is possible that every pair contains only 1 detected prime. In other words, the Twin Prime Conjecture may be true but the sieve method won't work to prove it. Some other idea is needed.

Whether you think the ultimate limit is 246, or 12, or 6, or 2, the lesson from Zhang and Maynard's work seems clear: Don't count number theorists out.

"Who knows what the next breakthrough will be?" asks Granville. "I'm very happy with this result, which is one of the great theorems in the history of analytic number theory. I would never have expected it in my lifetime."

"Who knows what the next breakthrough will be?" asks Granville. "I'm very happy with this result, which is one of the great theorems in the history of analytic number theory. I would never have expected it in my lifetime."

Computational Wizardry. *Computer scientists have long explored ways of ensuring that computational "wizards" give honest answers. A new theory of "rational proofs" dispenses with lie detection in favor of removing any incentive a wizard might have to lie.*

The Truth Shall Set Your Fee

Barry Cipra

WHAT MAKES YOU THINK THAT anything you're about to read is true? How much would you pay to be sure? And how sure do you need to be?

Economists have long known that honesty and trust are cornerstones of commerce. Those abstract qualities are sorely tested in the anonymous, ephemeral realm of cyberspace, especially when it comes to online merchants who peddle information instead of mere tangible goods. This is serious because we nowadays depend on computation—the engine of information—to keep our real world running reliably. As more and more digital processing takes place in "the cloud," offsourced to unknown agents of unproved reliability, it is more and more important to develop protocols for vouchsafing the results of computations we don't—or can't—perform ourselves.

The trick is, these protocols not only have to work, they need to be cheap.

Theoretical computer scientists are forging a framework for just such protocols. One recent development takes a page from economics itself: Silvio Micali and Pablo Azar at MIT have come up with a theory of "rational proofs," in which the seller of information (colorfully called Merlin) can maximize his profit only by telling the truth—indeed, will likely lose money if he lies. As long as Merlin behaves rationally, motivated by the chance to make money (as opposed to mere mischief), the information he provides can be trusted by the buyer (dubbed Arthur—the fanciful names suggesting a wealthy king who seeks advise from a computational wizard).

The Merlin-Arthur underpinnings for the theory of rational proofs go back to the 1980s, when computer scientists, including Micali, began exploring what are called interactive proofs. Interactive proofs belong to the theory of computational complexity, which studies the distinctions between things that are "easy" to compute and those that are apparently "hard."

The formal definitions of "easy" and "hard" are surprisingly technical, but two examples convey the gist: multiplication and factorization. It's not all that hard to multiply two numbers, say 1021×7121, even if you don't have a pocket calculator—you just multiply each digit of one number by each digit of the other, add up the results with appropriate "carries" and hope you don't make any mistakes. On the other hand, *factoring* a number, such as 7,268,407, can be a bear, even if you *do* have a pocket calculator. (If you don't believe me, give it a try.) Indeed, the difference in difficulty between multiplying and factoring is at the heart of a widely used system for secure communication known as the RSA algorithm.

Factorization demonstrates another crucial wrinkle in the theory of computational complexity: Even if factors are hard to find, they are, if found, easy to verify. That's not true of every hard problem, but it's true for a great many. These problems belong to a class that computer scientists label NP. The "easy" problems belong to a class labeled P. The "P" in each label stands for "polynomial time," which means that the amount of computation required to solve a problem (in P) or verify a solution (in NP) grows at most polynomially in the size of the problem. The "N" in NP stands for "non-deterministic"; roughly speaking, it means you're always free to *guess* an answer, and if you're lucky the polynomial-time verification will show you guessed correctly.

Over the last fifty years, computer scientists have developed a taxonomy of complexity classes that would boggle Linnaeus. The theory bristles with acronyms and subtle distinctions. And much as biologists occasionally discover that two branches in the tree of life are more closely related than they first appear, complexity theorists have repeatedly found surprising kinships when they sequence the DNA of computation.

Interactive proofs have provided some of the biggest surprises. The essence of interactivity is for Merlin to convince Arthur that the result of a computation he has performed is correct not by providing all the details, which may not be possible for Arthur to check, but by giving answers to some ancillary questions, the correctness of which Arthur *can* check. The wonder of it is that interactive proofs turn out to be amazingly powerful; through their use, Arthur can be fully confident Merlin isn't trying to pull a fast one.

Again, a simple example conveys the gist. The example is known as the Graph Isomorphism Problem: Given two graphs—collections of points (called vertices), pairs of which are connected by curves (called edges)—is there a way to label the vertices of each graph so that any time two vertices in one graph are connected by an edge, the "same" two vertices in the other graph are also connected by an edge, and vice versa? This seemingly simple problem is notoriously difficult; the number of different labelings grows exponentially with the number of vertices (to be precise, there are $n!$ ways to label a graph with n vertices), so a straightforward trial-and-error approach is

Pablo Azar and Silvio Micali. *(Photo courtesy of Lisa Liem.)*

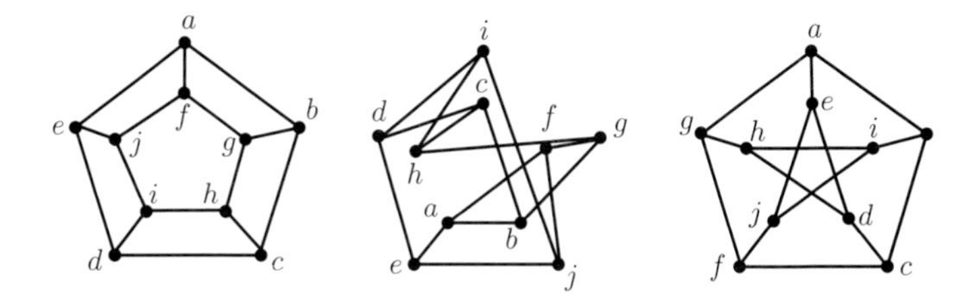

Figure 1. *One of these two graphs is not like the others. Graphs with ten vertices are easy to eyeball; graphs with ten thousand, or ten million, not so much.*

beyond the computational power of anyone but a wizard (see Figure 1).

If Merlin wants to convince Arthur that two graphs *are* isomorphic, he can do so by labeling them and letting Arthur check the labeling, edge by edge. But convincing Arthur two graphs are *not* isomorphic is trickier, and here's where interactivity helps. Let Arthur secretly select one of the two graphs, scramble it up to look completely different, and then ask Merlin which of the two original graphs it's isomorphic to. If the original two graphs are in fact isomorphic—that is, if Merlin is lying—then he can do no better than guess which one Arthur secretly selected. But if the graphs aren't isomorphic, then Merlin's computational wizardry will allow him to answer correctly. Moreover, he can do so no matter how many times the test is repeated.

There are two key features in interactive proofs. First, the "prover," Merlin, is assumed to have unbounded computing power, whereas the "verifier," Arthur, is limited to polynomial-time computations. Second, Arthur cannot rule out the possibility that Merlin got extremely lucky and gave correct answers during the interaction even though he lied about the actual result; but he can make that possibility exponentially remote, while staying within his polynomial-time budget. In the Graph (non-)Isomorphism Problem, each repetition of the test cuts a deceitful Merlin's odds of escaping detection in half. What wizard would risk his neck with a lie if he knows it'll be put to the test a hundred times?

In a full-fledged interactive proof, Arthur's line of questioning can be more byzantine, with questions probing not just Merlin's answer to the original problem, but also aspects of his answers to previous questions. Much as a clever lawyer reacts to potential slips in testimony, Arthur tailors his questions to amplify inconsistencies. The only constraint is that the interaction must stay within Arthur's polynomial-time budget.

The class of problems that are amenable to interactive proofs was given the acronym IP. In 1992, Adi Shamir at the Weizmann Institute of Science in Rehovot, Israel, showed that IP is, in fact, identical with a class of problems computer scientists had already given the name PSPACE. (Shamir is the "S" in "RSA" cryptography, the other two being Ronald Rivest at M.I.T. and Leonard Adleman at the University of Southern California.) Roughly speaking, PSPACE is a class of problems that Arthur himself could solve if he could somehow freeze time while his computer runs. Somewhat more technically, a PSPACE problem

Brier's formula is what's called a proper scoring rule. Over the long run, it gives the highest value to the forecaster who most accurately knows—and reports—the true chance of rain. In particular, it discourages anyone from shading their best estimate.

can be solved on a computer whose size (in particular, whose memory) is bounded by a power of the size of the problem.

Part of what makes interactive proofs able to solve PSPACE problems is that the number of rounds can grow polynomially with the size of the problem. But that's a limitation on their practicality. Simply put, Arthur may not have the time or patience to scramble a graph a hundred times to be sure Merlin is on the up and up; he may want to limit himself to a couple of questions. But he still wants confidence in the result.

Enter rational proofs.

Micali and Azar's theory aims at reducing the amount of interaction to as little as a single round. The key new wrinkle is a randomized payoff: At the end of the interaction, Arthur tosses some coins and then pays Merlin an amount that depends partly on the result of the coin tosses and partly on the answers Merlin gave during the interaction. If Merlin is rational, he will have provided answers that maximize the expected value of his reward. What Micali and Azar have shown is that payoff functions can be devised that maximize that expected value only if Merlin's answers are honest.

To do so, they reached back to an idea first proposed in 1950, in, of all places, a paper on weather forecasting.

Glenn Brier, a statistician with the U.S. Weather Bureau, published a paper in the *Monthly Weather Review*, "Verification of Forecasts Expressed in Terms of Probability." The problem is familiar to anyone who's ever wondered what it means when the weatherman says there's a twenty percent chance of rain. Brier's solution was to keep track of things with a simple sum over a set of forecasts: each time the forecast says the chance of rain is p and it actually rains, add p^2, while if it doesn't rain, add $(1-p)^2$. (And, of course, it makes sense to divide by the number of forecasts under consideration, to produce an average.)

Brier's formula is what's called a proper scoring rule. Over the long run, it gives the highest value to the forecaster who most accurately knows—and reports—the true chance of rain. In particular, it discourages anyone from shading their best estimate.

To apply this in a computing context, Micali and Azar fixed on a canonical problem in complexity theory that goes by the name #SAT, which belongs to a class called #P. The #SAT problem is to count the total number of True/False assignments for a set of logical variables that make a given logical expression true (see Figure 2). This turns out be about as hard as a counting problem can be. (The related "decision" problem, which simply asks if there are *any* True/False assignments that satisfy the given expression, belongs to NP. It is the original "NP-complete" problem. The counting version is #P-complete.) So if Arthur has a complicated logical expression with n variables, for which there are 2^n different True/False assignments in all, and Merlin tells him that exactly N of those assignments make the expression true, how should Arthur reward him? The answer is to treat Merlin's answer as a forecast: If you now pick True/False assignments for the n variables at random, Merlin's answer is, in effect, saying there's a $p = N/2^n$ chance of "rain." According to the Brier scoring rule, Arthur should pay Merlin a base amount of $p(1-p)$ and then a "bonus" of p if the random assignment turns out to make the expression true and $1-p$ if it makes it false (see Box, **"Great Expectations,"** page 34).

PQRS	$(P \vee$	$Q \vee$	$R) \wedge$	$(\neg P \vee$	$R \vee$	$S) \wedge$	$(Q \vee$	$\neg R \vee$	$\neg S) \wedge$	$(\neg P \vee$	$\neg Q \vee$	$\neg S)$
TTTF	T	T	T	–	T	–	T	–	T	–	–	T
TFTF	T	–	T	–	T	–	–	–	T	–	T	T
TFFT	T	–	–	–	T	T	–	T	–	–	T	–
FTTT	–	T	T	T	T	T	T	–	–	T	–	–
FTTF	–	T	T	T	T	–	T	–	T	T	–	T
FTFT	–	T	–	T	–	T	T	T	–	T	–	–
FTFF	–	T	–	T	–	–	T	T	T	T	–	T
FFTF	–	–	T	T	T	–	–	–	T	T	T	T

Figure 2. *Even a small #SAT problem—counting the number of True/False assignments that "satisfy" a logical expression (i.e, make it true)—can be a daunting task. Here exactly* 8 *out of* 16 *assignments for four variables satisfy the expression* $(P \vee Q \vee R) \wedge (\neg P \vee R \vee S) \wedge (Q \vee \neg R \vee \neg S) \wedge (\neg P \vee \neg Q \vee \neg S)$. *Or is there a mistake in the count?*

Micali and Azar parlayed #SAT into a proof that a single-round rational proof is capable of guaranteeing Merlin's answers to any problem in the class #P. Multi-round rational proofs, they showed, are capable of even more. Indeed, if the number of rounds is limited only by Arthur's polynomial-time budget, rational proofs are equivalent to interactive proofs. (Given a polynomial-time budget, rational proofs are at least as powerful as interactive proofs because Arthur can simply pay Merlin 1 if an interactive proof convinces him and 0 if it doesn't. The tricky part is to show that rational proofs confer no extra power.) But even when the number of rounds is fixed in advance, rational proofs apply to a vast number of computational problems (see Figure 3, next page).

The main appeal of rational proofs is that they reduce the amount of back and forth between a buyer and seller. But they still often require the buyer to do a lot of computation. And they also require the seller to be sensitive to exponentially small reductions in his expected reward. More recently, Micali and Azar have zeroed in on what they call super-efficient rational proofs, which address both these drawbacks.

In a super-efficient rational proof, Arthur never does more than a logarithmic amount of computation (beyond what it takes to specify the problem itself). This more closely models the real-world disparity between consumers of information, who might be using only a tablet or a smartphone, and producers, who potentially have server farms of supercomputers at their disposal. The logarithmic restriction puts its own limit on the types of problems that can be approached—#SAT, for example, provides answers that themselves are linear, not logarithmic, in the size of the problem—but super-efficient rational proofs still cover a lot of computational territory even if additionally restricted to a single round.

Micali and Azar's super-efficient approach also ensures that a malicious (or incompetent) Merlin will suffer polynomial-size losses to his expected income for giving wrong answers. To be sure, the original rational-proof theory can do this as well, but only if Arthur offers rewards that scale exponentially in the size of the problem. The newer theory shows that such penalties are

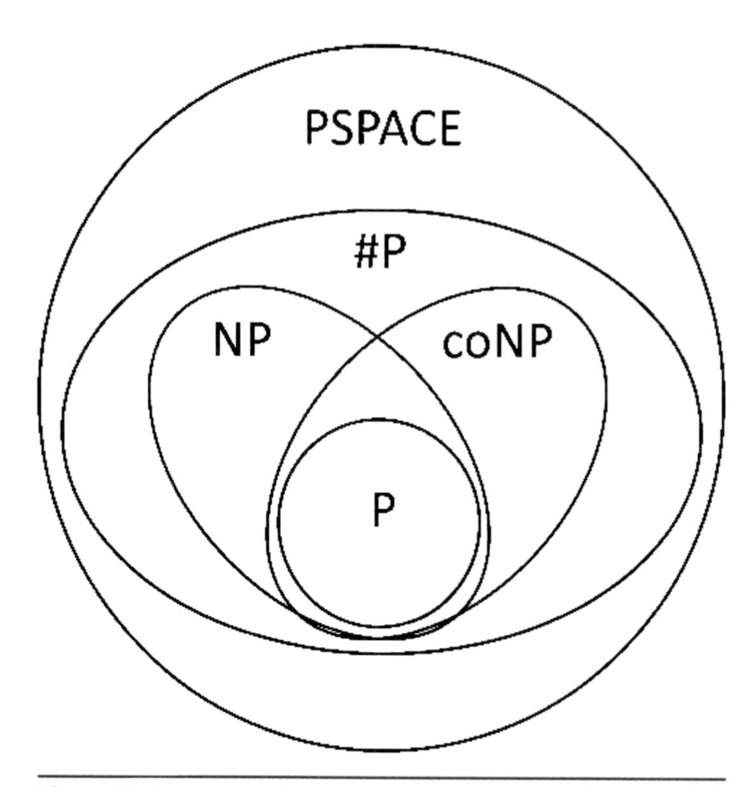

Figure 3. *As computational power grows to wizard proportions, the class of problems that can be solved broadens, from P to PSPACE, and even beyond. The relationships here are a blend of "decision" problems and "counting" problems. To be sure, theorists have not proved all of these classes are actually distinct, but it is widely believed they are. (Figure courtesy of Jing Chen.)*

possible with a polynomial-sized reward budget. (A polynomial-size reward can be computed in logarithmic time: It only takes a few seconds to write a check for a hundred dollars, and only a few more seconds to make it out for a million.)

The theory of rational proofs is still exploring the limits of what it can do. One of the open questions has only recently been solved. In their first paper, Micali and Azar asked whether Arthur could benefit by employing *multiple* Merlins. It turns out he can. Researchers Jing Chen, Samuel McCauley, and Shikha Singh at Stony Brook University have shown that multiple Merlins are strictly more powerful than one: They enable Arthur to confidently purchase solutions to an enormous class of computational problems (see Figure 4). In possibly the biggest surprise, that full power is achieved with just two Merlins and five rounds of interaction. Where the theory goes next may take its own Merlin to determine.

Great Expectations

Suppose Merlin knows the true probability of "rain" is p, but decides to report a probability x. If he does so, he knows he'll be paid a base amount of $x(1-x)$ regardless, and a "bonus" of

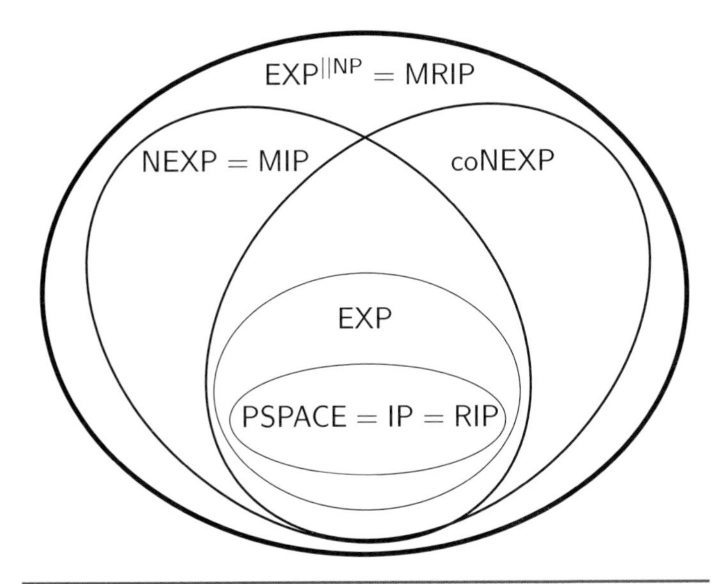

Figure 4. *When multiple Merlins enter the mix, the class of calculations that Arthur can rely on extends far beyond what he can trust when paying a single wizard. Amazingly, the entire extension is achieved with only two Merlins and five rounds of interaction. (Figure from "Rational Proofs with Multiple Provers," by Jing Chen, Samuel McCauley, and Shikha Singh, available at http://arxiv.org/abs/1504.08361.)*

x if it does indeed rain and $1 - x$ if it doesn't, so he knows his expected reward will be $x(1 - x) + px + (1 - p)(1 - x)$. Being a master of algebra as well as weather forecasting, he realizes this is equal to $1 - p + p^2 - (x - p)^2$. It doesn't require any wizardry to see that if the reported value x deviates from the true value p, his expected reward will *decrease* by a positive amount. So Merlin has every incentive to report the true value p. Arthur can count on this, even though he isn't privy to Merlin's thinking.

In general, Merlin knows that at the end of a rational proof, there will be a random challenge to the information he has provided and he'll be rewarded according to the result of the challenge. Merlin can anticipate the probability distribution of the challenge, so he can tailor his answers to maximize his expected reward. Micali and Azar's theory shows that, for a large class of computational problems, Arthur can structure the interaction, the challenge, and the reward so that Merlin's smartest move is to answer honestly.

The computed reward can, of course, be scaled according to the size of the problem—pennies for a small problem, dollars for a bigger one, and so on. Doing so is of practical importance since any real-world Merlin would have *some* costs to cover when he does a computation. It also opens a new line of analysis, since a cost-conscious Merlin could conceivably realize he's better off doing cheap calculations that bring him close to the maximum expected reward than costlier ones that actually achieve it. Micali and Azar have found ways around this too, with what they call super-efficient rational proofs.

Tipping Point. *Beginning in the 1980s, many Caribbean coral reefs were overgrown by algae. Mathematical models explain this phenomenon as a "tipping point," in which a gradual change to the parameters of a system (in this case, the ability of grazers such as sea urchins to keep up with the algae) causes the system to switch abruptly from one stable state (healthy coral) to a different one (overgrown coral). (Photo courtesy of Craig Quirolo, Reef Relief. [Photo 15 in Reef Relief Image Archive 12: Orange Bay shallow.])*

Climate Past, Present, and Future

Dana Mackenzie

BY NOW IT'S BEGINNING TO SOUND LIKE A BROKEN record. 1998: the warmest year in recorded history. 2005: warmest year in recorded history. 2010: warmest year in recorded history. 2014: warmest year in recorded history (see Figure 1).

Climate change is happening, and it is one of the most-discussed scientific problems of the twenty-first century. But beyond the fact that the climate is changing, little else is certain. Nobody knows how much change is inevitable, how much can still be averted, or which of the many climate systems are near to a "tipping point" that will lead to irreparable change. Will the ice sheets melt? Will the oceans rise, and by how much? Will the Atlantic Ocean "conveyor belt" slow down or shut off?

Climate scientists try to answer such questions by running very large climate simulations on the world's fastest computers, using equations that describe the physics of fluid flow and heat transport. "The complex models are great in the sense that they try to capture the physics. However, the process-level un-

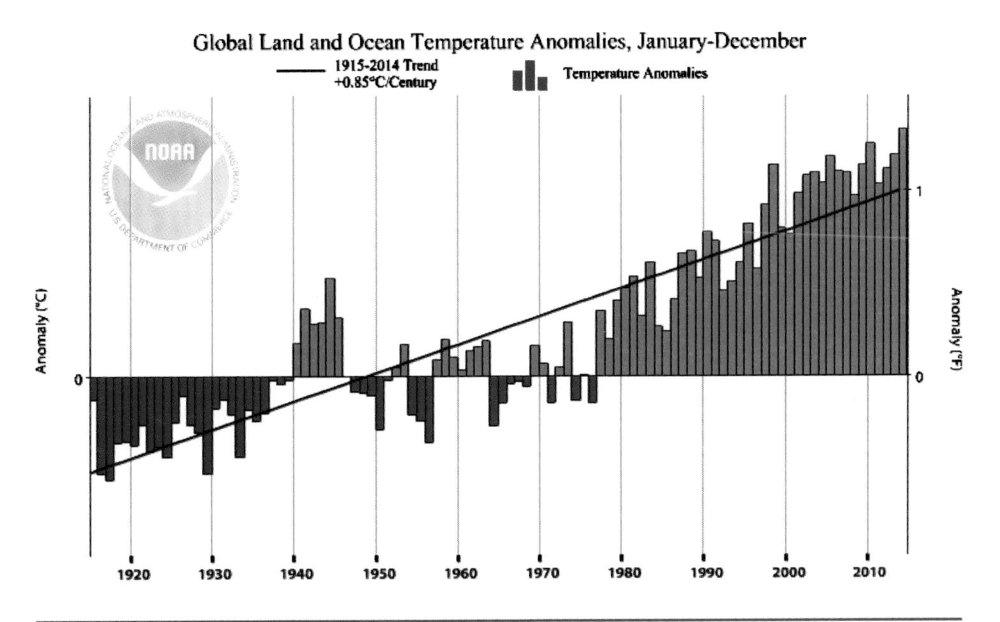

Figure 1. *Over the last 100 years, worldwide temperatures, averaging together both land and sea, have increased by 0.85° C. (1.53° F). This graph shows the linear trend for the years 1915–2015, superimposed on the observed temperature "anomalies." Zero anomaly represents the average temperature for the twentieth century, and 1° represents 1 degree (Fahrenheit) above the global average for the twentieth century. Records were set in 1998, 2005, 2010, and 2014. (Figure courtesy of NOAA/NCEI.)*

...[H]igher life on Earth was nearly done in by the humblest of microbes, a single-celled organism called *Methanosarcina.* The story behind this hypothesis reads like a whodunit, with the first and most important clue coming from a mathematical anomaly that most climate scientists might have missed.

derstanding is quite difficult to get at," says Graham Feingold, a cloud physicist at the National Oceanic and Atmospheric Administration. "The prevailing culture [in climate studies] is still one of adding detail, building as much physics as you can into the system. But there's a bit of a counterculture of people who try to build simpler models."

That's where mathematicians come in. The natural instinct of mathematicians is to simplify. To determine whether a tipping point will occur in a natural system, they first ask what is the *simplest* system that exhibits such a tipping point, and try to calculate the tipping point in that model. However, it is important not to fall in love with that model. The goal must be to transfer the "process-level understanding" gained from a simple model to more complicated ones. "If mathematicians want to contribute, they need to move away from the comfort zone of solving a couple of ordinary differential equations," says Sean Crowell, a mathematician at the National Weather Center in Norman, Oklahoma.

Here are three stories of how mathematicians are contributing to the search for answers about our changing climate.

Climate Past: How the Microbes (Nearly) Won

No matter how much the climate may warm up over the next century, Earth has seen worse times. It is almost impossible to imagine a future more grim than the end-Permian extinction. This geological event, also called "the Great Dying," was the narrowest bottleneck that life on Earth ever passed through. About 252 million years ago, over a period that may have been as short as 12 thousand years, more than 90 percent of the animal species on Earth became extinct. Trilobites, which had survived for hundreds of millions of years, disappeared and were never seen again. Even insects, which were left unscathed by other mass extinctions, were ravaged by the Great Dying (see Figure 2).

If we could visit Earth right after the mass extinction, it would seem like a different planet—an arid, treeless place. The low oxygen levels would leave us gasping for breath, and the carbon dioxide level would be three times higher than today. In fact, the atmosphere would reek of sewer gas, with large amounts of hydrogen sulfide and methane as well.

What could have wreaked such environmental havoc? According to Dan Rothman, a geophysicist from MIT, higher life on Earth was nearly done in by the humblest of microbes, a single-celled organism called *Methanosarcina.* The story behind this hypothesis reads like a whodunit, with the first and most important clue coming from a mathematical anomaly that most climate scientists might have missed.

There has been no shortage of theories about what could have caused the end-Permian extinction. Perhaps it was a meteorite impact, like the one that killed the dinosaurs. But to date, there is no smoking-gun evidence of a crater. In fact, the evidence points toward volcanism as a chief suspect, because the end of the Permian period coincides with the formation of the largest volcanic deposits on Earth, the Siberian traps.

It makes for a tantalizingly simple story. The Siberian rifts belched forth lava and gas for 10,000 years, creating a hell on Earth and releasing thousands of gigatons of carbon into the atmosphere. The greenhouse effect warmed the climate on land,

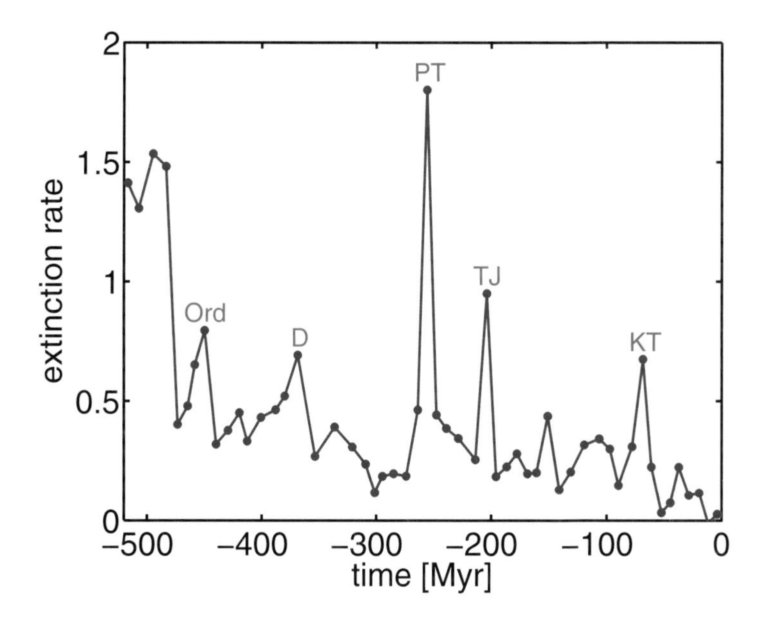

Figure 2. *Relative extinction rates, measured at the genus level, over the last 500 million years of Earth's history. Five "mass extinctions" are clearly visible, which took place between the Ordovician and Silurian Periods (Ord), during the late Devonian (D), between the Permian and Triassic (PT), between the Triassic and Jurassic (TJ), and between the Cretaceous and Tertiary (KT). (Figure from Extinction rates from J. Alroy,* Paleobiology, *vol. 40, pages 374–397. Figure courtesy of D. H. Rothman.)*

and acidification of the oceans killed off most of the organisms in the sea.

What could be wrong with this scenario? One mathematical detail. "In models of massive volcanism, there are holes punched in Earth's surface called vents," Rothman says. "With each decrease in overpressurization, we should see an increase in the time between formation of new vents." In other words, there should be a gradual slowdown in the injection of carbon into the atmosphere, similar to the gradual cool-down of a hot cup of coffee.

But that is the opposite of what Rothman saw in two rock deposits laid down during the end-Permian extinction, one in Meishan, China and the other in the Austrian Alps. Instead of a gradually tapering growth in the amount of carbon, he saw a runaway increase, called *superexponential* growth. Around 252.5 million years ago, the amount of inorganic carbon in the ocean began galloping toward a vertical asymptote, which it reached 20 thousand to 30 thousand years later (see Figure 3, next page).

The distinction between sublinear and superexponential growth may sound like something only a mathematician could get excited about. But in fact, it provides an important clue to what went wrong. Sublinear growth suggests a physical cooling-down process. Exponential growth suggests... biology. Some new organism was taking off, unchecked by competitors or limited resources. "I had a hunch that it might be related to some changes in the microbial biosphere, related

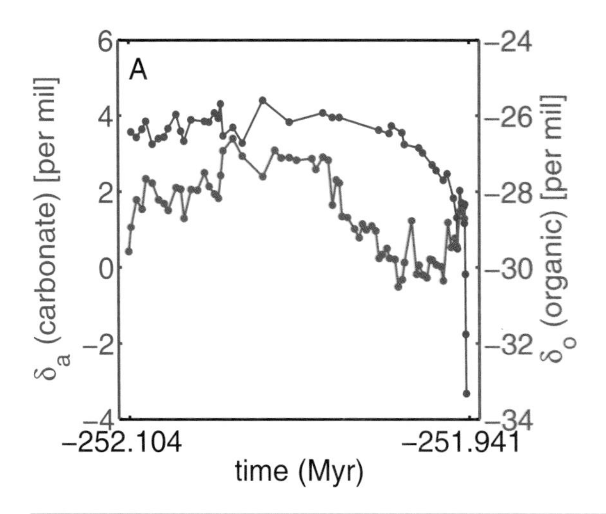

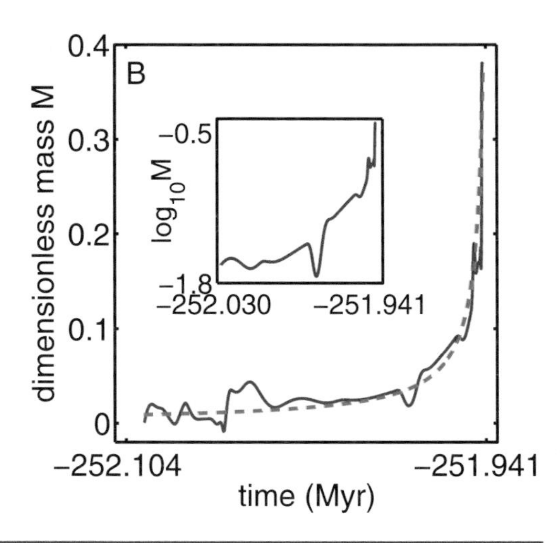

Figure 3. *Two graphs that suggested a biological explanation for the Permian extinction. Left, measured abundances of organic and inorganic carbon in end-Permian sedimentary rocks from Meishan, China. Right, inferred addition of isotopically light carbon to the global oceans. The rate of increase is superexponential, as shown by the inset, where the carbon curve is graphed in logarithmic coordinates. (Figure from Carbon isotopic data of part (a) from C.-Q. Cao et al.,* Earth and Planetary Science Letters, *vol. 281, pp. 188–201. Figure courtesy of D. H. Rothman.)*

to a new metabolism," Rothman says. "That was my working hypothesis."

If exponential growth suggests biology, SUPER-exponential growth suggests one more thing... doomsday.

"There's a famous, classic paper that appeared in *Science* magazine in 1960, by Heinz von Foerster, with the lovely title 'Doomsday: Friday, 13 November, A.D. 2026,'" Rothman says. In this paper, von Foerster and his co-authors fit the trend of world population to a superexponential growth curve, and found that the population would approach infinity on the date in the title.

Of course the title of von Foerster's paper was meant partly in jest, but there was also a serious point to it. Superexponential growth results from a tiny tweak to exponential growth. A system that grows exponentially obeys a differential equation like this one:

$$\frac{dA}{dt} = a_0 + a_1 A.$$

Here a_1 is assumed to be positive. This equation says that the rate of growth of A (say, the population of an organism) is a linear function of A. This is typical when there are no checks on the organism's growth, so that it reproduces with a constant birth rate.

Ordinarily, in the real world, populations cannot run amok forever. Eventually organisms start competing for resources, and this gives rise to an equation with a negative quadratic term representing the competition:

$$\frac{dA}{dt} = a_0 + a_1 A - a_2 A^2.$$

This kind of differential equation leads to logistic growth, where the population A will approach a finite limit, called the *carrying capacity*. Such growth laws are ubiquitous in natural systems.

But what if, instead of competing with each other, the organisms create a positive feedback, so that the more organisms there are, the more favorable the environment become for them and the faster they can reproduce? This sort of positive feedback would be modeled by a differential equation with a positive quadratic term:

$$\frac{dA}{dt} = a_0 + a_1 A + a_2 A^2.$$

Such an equation inevitably leads to a population that explodes to infinity in a finite time—the "doomsday" scenario envisioned by von Foerster.

Of course an infinite population can never actually happen. The system will have to break before it hits the asymptote. But this breaking can be a profoundly destructive process. Even if a species hits a superexponential growth curve only once or twice in Earth's history, that is one or two times too many. Rothman hypothesized that this is what happened at the end of the Permian period.

Dan Rothman *at Meishan, China. (Photo courtesy of Francis Ö. Lustwerk-Dudás.)*

At first, he says, it was a guess and nothing else: "I had no idea about the methanogens." But he consulted with Eric Alm and Alm's student Greg Fournier, who had recently studied a one-celled organism in the kingdom of archaea, called *Methanosarcina.* Fournier's work has shown that this organism had evolved a new, more efficient method of converting acetate to methane at some time in the past 450 million years. This could be the "new metabolism" that Rothman had hypothesized. Using computational genomics (itself a mathematical method—but outside the scope of this chapter) Fournier calculated the origin of *Methanosarcina* at 200 to 280 million years ago, an interval that plausibly brackets the end-Permian extinction.

Now Rothman had a suspect, and he could place the suspect at the right time. But could he place the suspect at the scene of the crime?

The whodunit analogy may be a little forced here (after all, the Great Dying was a global process), but in fact there was one more piece of evidence to come. "There was always a question in my mind, was there something about volcanism that would have made such an evolutionary event more likely to be successful?" says Rothman. "I asked [Alm and Fournier] if *Methanosarcina* requires any trace nutrients. They looked at me and said, nickel."

The last piece of the puzzle clicked into place. The Siberian traps contain the world's largest deposit of nickel. And, sure enough, when MIT graduate student Kate French analyzed the Meishan sediments, she found a spike in the amount of nickel at exactly the right time.

And so Rothman and his colleagues arrived at a more nuanced version of the volcanism story. The Siberian eruption was an accessory to the end-Permian extinction, but not the culprit. The eruption created a huge worldwide stockpile of carbon and nickel. Around the same time or a little earlier, a chance mutation created the new breed of methanogens. The nickel provided a nearly unlimited supply of an unusual element they needed to survive. They began pumping out methane—a greenhouse gas 25 times stronger than carbon dioxide. Temperatures went up and their ecological niche (or carrying capacity) expanded, enabling them to metabolize carbon faster and hop onto the superexponential express.

The result was doomsday. Massive amounts of carbon went into the oceans, acidifying them and killing off nineteen out of every twenty marine species. Another byproduct of the microbial metabolism, hydrogen sulfide, would have poisoned animal life on land. Then, because it had to, the superexponential express derailed—but not before it had utterly transformed the landscape.

Are there lessons for today in the Permian extinction? Yes. One lesson is that microbes are important. Another is that the carbon cycle on our planet is sensitive, and not to be tampered with lightly.

Finally, Rothman adds, "It's kind of important to say that the whole line of thought was enabled by mathematics." But it didn't stop with the mathematical equation. The real story is the way that mathematics provided a thread of evidence that bound together chemistry, biology, genomics, and earth science.

Climate Present: A Matter of Some Gravity

One of the most notable harbingers of climate change has been the melting of polar ice caps. Strangely, the Southern Hemisphere has behaved quite differently from the Northern Hemisphere. While the ice sheets in the Arctic Ocean and Green-

Christopher Harig and Frederik Simons. *(Photo courtesy of Frederik Simons.)*

land have been melting faster than anyone predicted, the area of sea ice in the Antarctic has increased modestly (about 1 percent per decade) since observations began in 1979. Is the amount of ice in the Antarctic truly increasing? If so, how can one reconcile this observation with increasing temperatures?

One of the most notable harbingers of climate change has been the melting of polar ice caps. Strangely, the Southern Hemisphere has behaved quite differently from the Northern Hemisphere.

According to Frederik Simons, a geophysicist at Princeton University, appearances can be deceiving. "Suppose that you're watching your weight," he says. "There are three ways that you can do it." One is to look in the mirror. This is the analogue of looking at Antarctica from space, using altimetry or photogrammetry. But an apparent change in volume does not necessarily translate to a change in mass. "You can't tell if you're gaining muscle or fat," Simons says, continuing the weight-loss analogy.

A more complicated approach is to watch your calories—all the food you take in, all the energy you expend. The climate analogue would be to measure the precipitation, ablation, and melting of all the ice in the continent—in essence, studying the "metabolism" of the ice sheet. Many geophysicists are working on this, but it's not easy because there are few weather stations on the frozen continent.

Finally, the gold standard for telling whether you have gained or lost weight is, of course, to step on the scales. That is the approach that Simons has been using. Since 2002, a satellite called the Gravity Recovery and Climate Experiment (GRACE) has measured the changes in Earth's gravitational field from orbit, and the idea is to work backward from the gravitational field to the mass distribution that generated it.

However, GRACE makes *global* measurements. NASA releases the data in the form of spherical harmonic coefficients. This approach is very much like digitally processing the sound of a drum into a fundamental frequency and overtones. The "strength" of the first spherical harmonic is the Earth's mass. The strength of the next is its dipole moment—the extent to which Earth is stretched out like a football or flattened like a bean bag. Then come quadrupole moments, octupole moments, and so on. The GRACE data go all the way up to 60-pole moments, which make it possible to resolve bulges or dimples in Earth's gravitational field as small as 330 kilometers in size.

To reconstruct Earth's gravitational field, GRACE measures 3721 moments in all. If you are dealing with only a piece of Earth's surface, like Antarctica, you should only need a piece of that information. But spherical harmonics don't come packaged that way. Every spherical harmonic integrates data from the entire globe, so you need measurements of the gravitational field over Hawaii to deduce the gravitational field over Antarctica.

This may sound absurd, but it's actually a common problem in climate science: how to scale a global model down to a regional one. Simons was certainly not the first person to look for a way to repackage the global GRACE data into Antarctica-shaped pieces. But he was the first to realize that the key mathematical tool had been discovered half a century earlier.

In the late 1950s, researchers at Bell Laboratories wanted to find the most efficient way to transmit information over a telephone. This information comes in two forms: frequency (the pitch of an audio signal) and time (the moment when it occurred). A mathematician named David Slepian derived the optimal way to encode information about a signal with a fixed bandwidth of frequencies and a fixed time interval. The so-called "Slepian functions" tailor the harmonics of an audio signal to a specific rectangular window in the time-frequency plane.

Though Slepian functions were designed with electronic communication in mind, they are equally applicable to the gravity problem. The main modification is that the spherical harmonics have to be tailored to an Antarctica-shaped, rather than rectangular window. The modified Slepian functions filter out the information in the spherical harmonics that is irrelevant to Antarctica, and reduce the enormous orchestra of 3721 harmonics to a much more manageable band of 100. With this simpler set of functions, Simons could deduce not only the *total* mass of ice lost in Antarctica, but also where it is being lost (see Figures 4 and 5).

Simons and his postdoctoral colleague Christopher Harig calculated that Antarctica has been losing ice since 2002. The loss has been accelerating, and the greatest losses are occurring in the West Antarctic Ice Sheet. This region was already known to be the one at greatest risk for a complete disintegration of

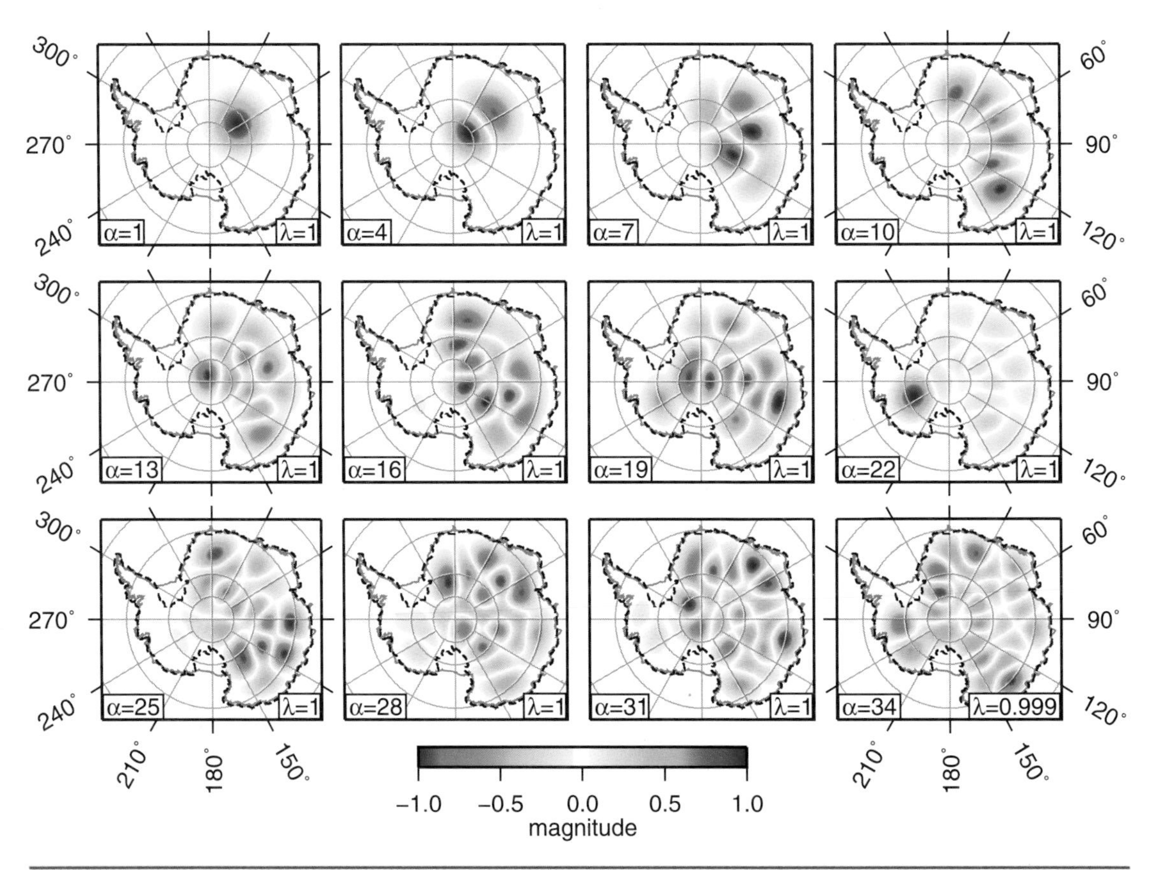

Figure 4. *Slepian functions for Antarctica. Alpha indexes the function number; the functions progress from simpler to more complicated as alpha increases. These functions can be used to decompose global gravity-field measurements into the components that are most informative about changes in Antarctica and its subregions. Function 1, for instance, is most sensitive to variations in the center of the continent, while function 22 picks up variations that are localized to West Antarctica. (Figure courtesy of Frederik Simons.)*

the ice sheet. At the same time, the East Antarctic has seen an increase in ice mass, probably due to increased precipitation, but the increase is only about half the amount lost in the West.

Other researchers have confirmed Simons and Harig's findings (using the "count the calories" method and the "look in the mirror" method). However, Simons prefers the rigor, the robustness, and the temporal and spatial precision of the Slepian-

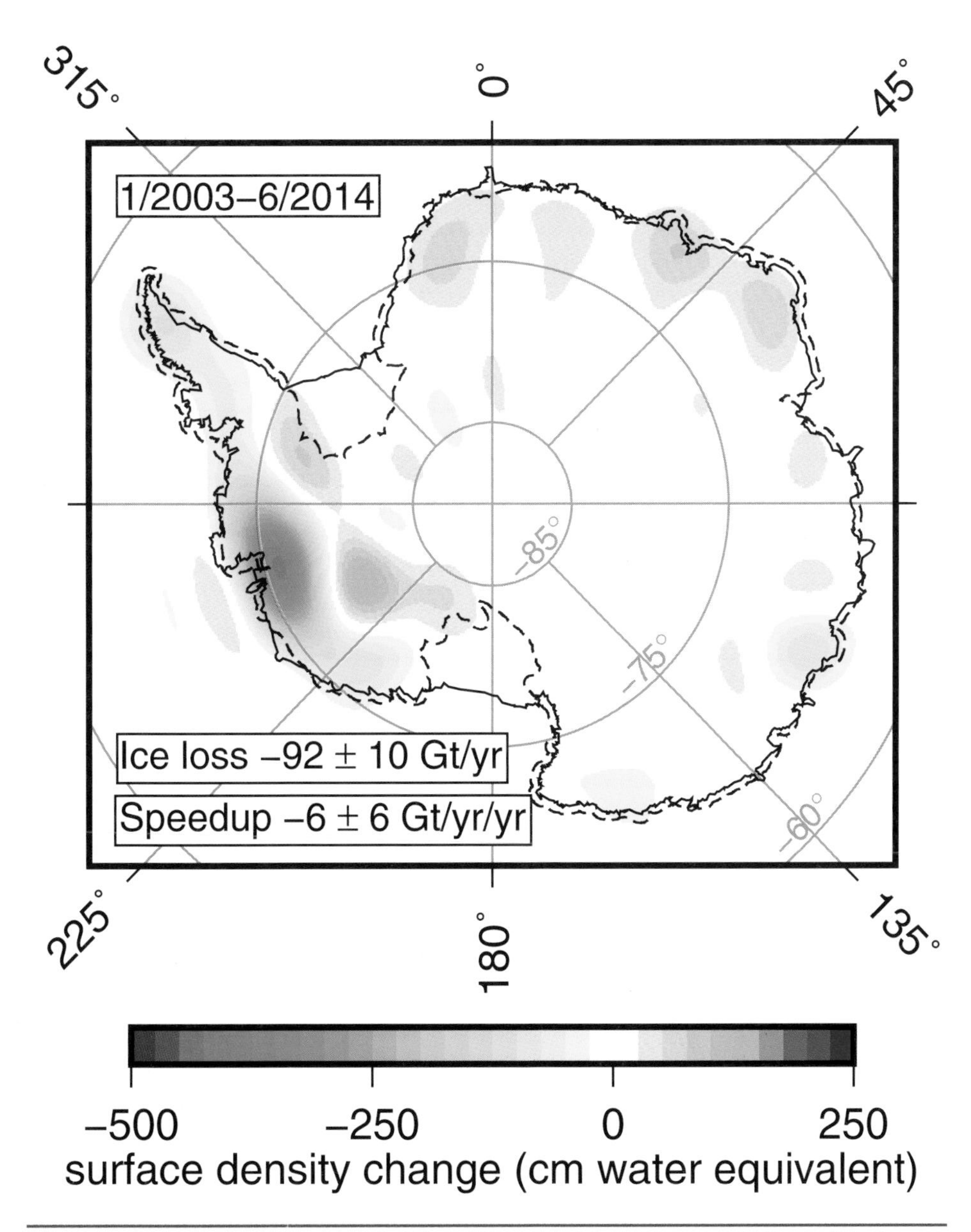

Figure 5. *Ice loss in Antarctica from 2003 to 2014, reconstructed from global gravity data filtered through the Slepian functions. Dramatic and accelerating ice loss in West Antarctica (red) is not compensated by gains in East Antarctica (blue). (Figure reprinted from "Accelerated West Antartic ice mass loss continues to outpace East Antartic gains," by Christopher Harig and Frederik J. Simons,* Earth and Planetary Science Letters, *with permission from Elsevier.)*

function approach. "What we see, we see with confidence," he says.

Climate Future: Tipping Points

In some ways climate change is a gradual process, whose effects are almost invisible on the time scale of a human life. If the temperature rises by 2 degrees over the course of a century, or sea levels rise by half a meter (both likely scenarios under the 2013 Fifth Assessment Report of the Intergovernmental Panel on Climate Change), most people would not even notice it.

However, there are at least two reasons why this gradual change can have far-from-gradual consequences. One is that as the climate changes, weather extremes—such as storms or droughts—become even more extreme, and these are precisely the events that cause the greatest damage and loss of life. The second concern is that gradual change can lead to a *tipping point*, beyond which the climate equilibrium can change suddenly and dramatically.

....[A]s the climate changes, weather extremes—such as storms or droughts—become even more extreme, and these are precisely the events that cause the greatest damage and loss of life.

Mathematicians have studied tipping points for decades, usually under a different name: *bifurcations*. A simple kind of bifurcation occurs in the differential equation $\frac{dx}{dt} = a - x^2$. This is a one-dimensional dynamical system, described by one state variable $x(t)$. The equation also has one parameter a, which we think of as a constant or as a slowly changing variable that is independent of x. If the parameter a is positive, then the system has two equilibrium states where $\frac{dx}{dt} = 0$, namely the points $x = \pm\sqrt{a}$. The positive equilibrium point is attracting and the negative one is repelling, which means that if x starts close enough to $\sqrt{a}$, $x(t)$ will move closer and closer to $\sqrt{a}$ as the time t moves on.

If $a = 0$ the two equilibrium points merge into one equilibrium point. Finally, if a is negative, the system has no equilibrium. In fact, this would be a doomsday scenario just like the ones described in the previous section, where $x(t)$ would crash to $-\infty$ in a finite time.

Now imagine varying the parameter a very slowly, so that the system has a chance to equilibrate. Then $x(t)$ will stay very close to $\sqrt{a(t)}$, right up to the moment when a hits 0: the tipping point. But if a decreases even the slightest bit further, $x(t)$ will have nowhere to go. In this way, a gradual change in the parameter a leads to a catastrophic change in the state $x(t)$.

The process described above, in which a stable and unstable equilibrium merge and annihilate one another, is called a *saddle-node bifurcation*. It is not the only kind of bifurcation, but it is the most common and the most basic. If we know the differential equation describing the dynamical system, we can in principle tell exactly where the tipping points are.

In practice, though, we don't always know the differential equations. And even if we did, solving for the tipping points would be far too complex a problem. For example, if we are studying the ocean, we are dealing not with one variable but with a huge number of them. We would like to describe the motion of every gallon of water in the ocean, and every gallon adds a new set of variables. Depending on how faithful you want your model to be, it may have thousands, millions, billions of variables. Finally, even if you could find the bifurcation points in your model, you have to recognize that the model is only an imperfect representation of reality.

For all the reasons above, we have to look for indirect methods of diagnosing a tipping point. In fact, dynamical systems behave in certain characteristic ways when they get close to a saddle-node bifurcation. One telltale sign is that the system starts responding more slowly to a disturbance; as its equilibrium point becomes less stable, the system takes longer to return to it. This slowing down can be detected as an increased correlation between earlier data points and later ones. Likewise, if there is spatial data, the spatial fluctuations will also become larger, and each point in the data will become correlated with a larger neighborhood of surrounding points. So increased autocorrelations (correlations between the same point at different times) or telecorrelations (correlations between distant points) are a warning sign: tipping point ahead.

One part of the climate system that may be sensitive to perturbations is the "conveyor belt" of ocean currents that bring warm tropical water north via the Gulf Stream. If this circulation were to shut down, northern Europe would lose a heat source, and a new Ice Age could ensue. The slowly varying parameter that could shut it down is the freshwater input into the Atlantic due to melting of Arctic glaciers and sea ice. Thus, ironically, global warming could lead to a tipping point that would trigger rapid global *cooling*, at least until the conveyor belt started up again. This scenario is considered unlikely to happen in the 21st century, but it has happened in the past and could happen again in the more distant future.

One part of the climate system that may be sensitive to perturbations is the "conveyor belt" of ocean currents that bring warm tropical water north via the Gulf Stream. If this circulation were to shut down, northern Europe would lose a heat source, and a new Ice Age could ensue.

Henk Dijkstra of the University of Utrecht and others have tested whether correlations can be used to detect tipping points in a model of the northern Atlantic Ocean. In a computer simulation, Dijkstra set up a *climate network*, a collection of observation points in the Atlantic Ocean (see Figure 6, next page). Any two points in the network are considered to be connected if their correlations exceed a threshold value, such as 0.7. The simulation was performed first without glacial melting and then again with a freshwater influx that gradually increased over two millennia to an amount equivalent to 10 Amazon Rivers.

Without glacial melting, the conveyor belt of course remained undisturbed. With glacial melting, it reached a tipping point and collapsed at about the 874th year of the simulation. But most importantly, the climate network gave an early warning signal of increasing connectivity that started at year 738 and lasted for more than 50 years—a strong enough and early enough signal (one might hope) that governments could get together and take some preventative action.

"Ah, but it's only a computer simulation," you might say. That's true. And it's also true that the early warning system is *based on the assumption* that the ocean circulation has a simple saddle-node bifurcation. But this is why it's so important to document the tipping point through an increasingly complex hierarchy of models. Dijkstra began with a simple two-dimensional system where he could calculate the bifurcation explicitly. Then he proceeded to a more complicated system where the bifurcation could not be calculated, but could still be verified to exist around the same parameter values. Knowing that, he could look for early warning signals that diagnose the tipping point not from the (poorly known) parameters but from the (accurately measured) climate observations. These early warning signals could then be generalized to the most sophisticated climate models.

While some climate researchers look for early warning signs, others are scrutinizing what the word "tipping point" actually means. "The tipping point concept has been accepted very well in climate science, but only after they stopped calling them bifurcations and started calling them tipping points," Dijkstra says. "The word is a little bit more vague and gets everyone interested."

"It isn't very well defined, but maybe one shouldn't constrain it too much," says Peter Ashwin of the University of Exeter. There are other types of tipping that do not fit into the classical picture of bifurcations. One of them is *noise-induced tipping*. Here a little bit of random variation is added to a traditional, deterministic dynamical system. If the deterministic system has two equilibrium states (say, sunny and rainy) with a small enough threshold between them, the random fluctuations might be enough to make the system jump from one state to the other.

Ashwin (along with his colleagues Sebastian Wieczorek of University College Cork, Catherine Luke and Peter Cox from the University of Exeter) recently identified a third kind of tipping,

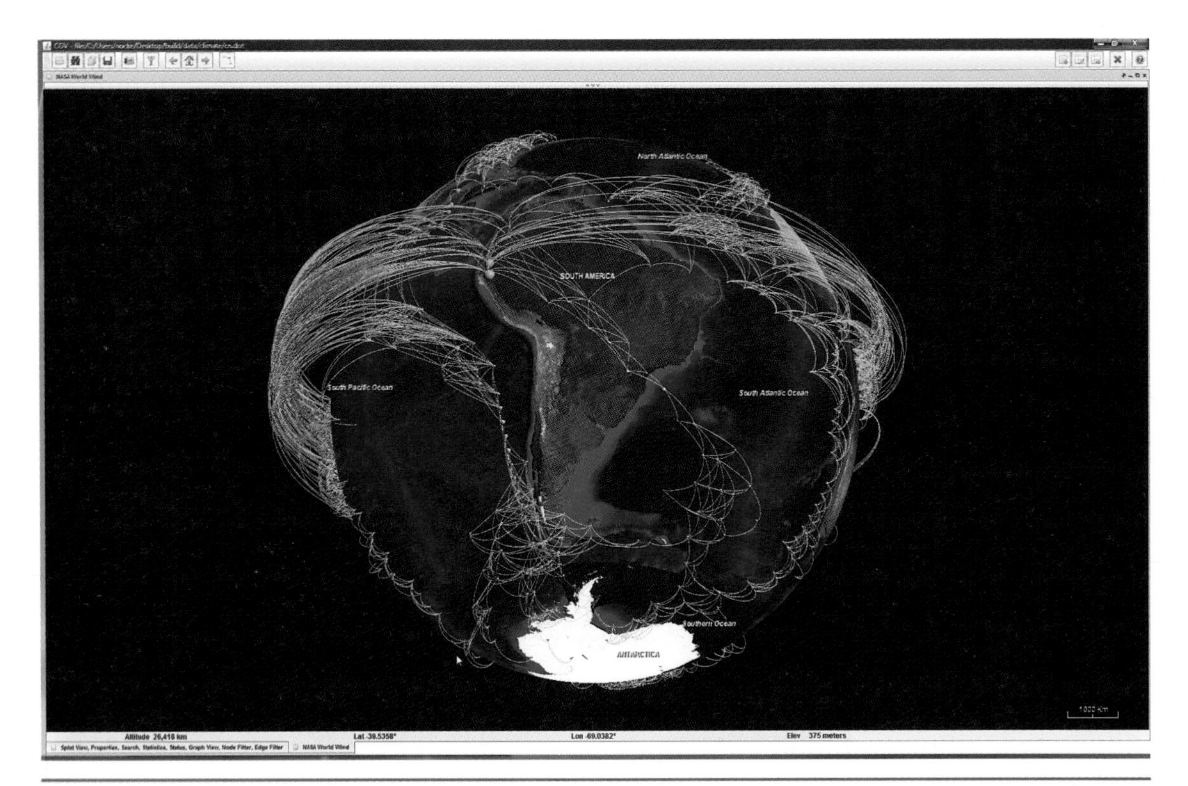

Figure 6. *A climate network developed at the Potsdam Institute for Climate Impact Research (PIK). Nodes are points where a time series of surface air temperature measurements is available. Links connect points whose time series are particularly strongly correlated. Note, for instance, strong correlations in the South Pacific due to the El Nino phenomenon. It is believed that one sign of a climate tipping point would be an increase in the length and density of these connections. (Figure created with the network visualization tool CGV (http://www.informatik.uni-rostock.de/~ct/software/CGV/CGV.html.) Courtesy of Potsdam Institute for Climate Impact Research.)*

called *rate-induced tipping*. It occurs when the parameters of a dynamical system are varied too quickly to allow it to equilibrate. In the classic picture of a saddle-node bifurcation, the parameter is assumed to change very slowly, so that the system has a chance to evolve along its stable equilibrium curve. But in rate-induced tipping, the system lags behind the equilibrium curve. It can eventually get so far away that it gets attracted to a different equilibrium point. It can even be attracted to a supposedly unstable equilibrium, because the instability does not have enough time to manifest itself. This kind of "impossible solution," like a chair balancing on one leg, behaves like a false attractor, and it often serves as an intermediate stage as the system moves from one stable equilibrium to another (see Figure 7, next page).

One example is colorfully known as the "compost-bomb instability." Climate scientists have hypothesized that an increase in temperature could cause the world's peatlands to overheat, like a compost pile. Indeed, Russia had major peat fires in 2010. Ashwin's group modeled the peatlands with a system of two equations, in which the average atmospheric temperature is the slowly varying parameter. In that model, they found that a tipping point is reached, leading to spontaneous combustion, if the temperature of the atmosphere increases faster than 9 degrees per century. (This is yet to be confirmed in a full-scale model.)

Rate-induced tipping may actually be common in ecological systems. For example, it may explain why coral reefs in the Caribbean Sea were overgrown by algae beginning in the 1980s (see **Tipping Point**, page 36). The algae are normally grazed on by herbivores, such as sea urchins. According to a model developed by Marten Scheffer of Wageningen University in the Netherlands, the reef has two stable equilibrium states, one "herbivore-dominated" and one "algae-dominated," with an unstable equilibrium (or tipping point) between them. If a sufficiently rapid pulse of nutrients enters the ecosystem, for example runoff water from land during a hurricane, the system shoots past the unstable equilibrium and enters the algae-dominated state, where it remains even after the temporary infusion of nutrients is over. By contrast, if the same amount of nutrients is released gradually, the system stays in the herbivore-dominated zone and the reef recovers. Ashwin considers this to be a case of rate-induced tipping because the *rate of release* of nutrients, not the absolute amount, determines whether the tipping occurs.

Scheffer and three colleagues hypothesized that the collapse of the coral reefs came about because of overfishing, which left only one abundant herbivore species (the aforementioned sea urchins). When a pathogen sickened them, they were no longer able to keep up with the algal blooms.

....[T]he collapse of the coral reefs came about because of overfishing, which left only one abundant herbivore species (the aforementioned sea urchins). When a pathogen sickened them, they were no longer able to keep up with the algal blooms.

Bridging the Gaps

As these examples show, mathematicians have a major role to play, not only in building climate models but also in formulating the right questions to ask about them. For a long time, though, their voices were not very loud.

"Historically, there has been a long tradition of mathematical physics, but not so much of mathematical geophysics," says

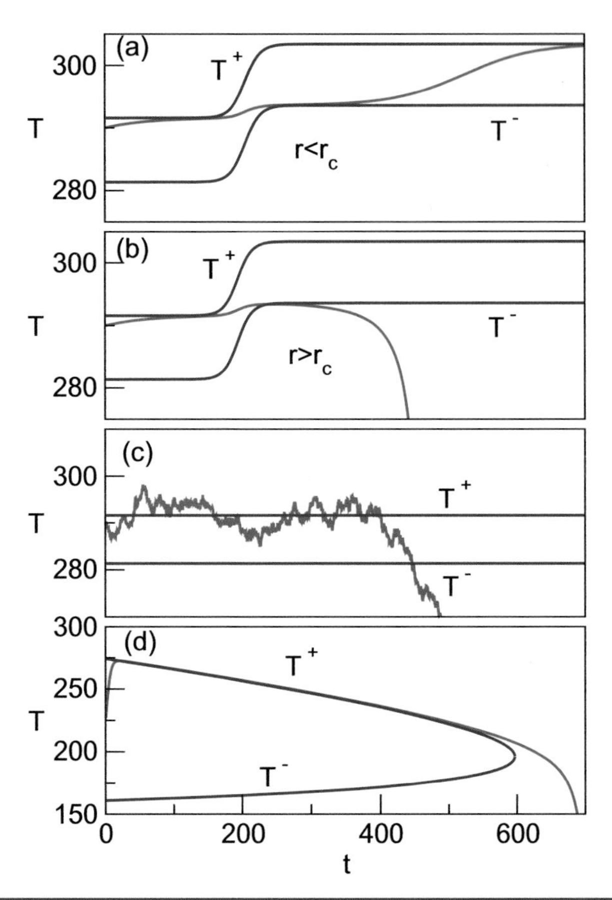

Figure 7. *In a recent paper, Peter Ashwin, Sebastian Wiezcorek, Renato Vitolo and Peter Cox showed that all three kinds of tipping points (bifurcations, noise-induced tipping, and rate-induced tipping) can exist in the same model. The model, proposed by Klaus Fraedrich (1979) and Alphonso Sutera (1981), describes the change in temperature (T) on a fictional planet as a function of solar radiation, thermal capacity of the ocean and ice-albedo feedbacks. Sutera's version also includes a random term.*

Equilibrium temperatures T^+ (stable) and T^- (unstable) are plotted in blue, while actual temperatures are plotted in red. In (a) and (b), a change to the planet's ice-albedo feedback causes the equilbrium temperatures to increase. In scenario (a) the rate of change is slow enough for the planet to return to the new, higher stable equilibrium after briefly tracking the "wrong," unstable equilibrium. In (b), the rate of change is too great, the temperature fails to bounce back, and the world becomes a snowball. In (c), with random noise, a run of unusually cold years can throw the planet out of equilibrium, even without any change to the system's physical parameters. In (d), a change in the incoming solar radiation leads to a classic saddle-node bifurcation, where the planet is left with no equilibrium state except a snowball. (This is a modified version of a figure from "Tipping points in open systems," Phil. Trans. R. Soc. A *(2012)* **370**, *1166–1184. Courtesy of Peter Ashwin.)*

Ashwin. He thinks that "institutional barriers" have kept mathematicians and geoscientists apart.

The situation started changing in the late '00s. A workshop on the mathematics of climate change at the Mathematical Sciences Research Institute (MSRI) in 2008 brought together professional climate modelers with several dozen mathematicians for the first time. This workshop led to the proposal and funding of the Mathematics and Climate Research Network (MCRN), which is still active and has a website at www.mathclimate.org. MCRN provided both a forum and job opportunities to students and postdocs interested in climate issues. Sean Crowell, who was a beginning graduate student in 2008, says, "I was always concerned about stewardship of our planet, but I didn't think I'd have a chance to do it professionally. [MCRN] has perfectly united my values with my professional field." After getting his doctorate in mathematics at the University of Oklahoma, Crowell was hired for a postdoctoral position at the National Severe Storms Laboratory before moving to his current position at the National Weather Center. These opportunities might not have been open to a math Ph.D. a decade ago.

Christiane Rousseau. *(Photo courtesy of Christiane Rousseau.)*

If MCRN made the mathematics of climate change into a national concern, a single mathematician named Christiane Rousseau took it international. In 2009, she proposed the idea of a climate year to several Canadian mathematics institutes. They sent around a one-page proposal to their American counterparts. As Rousseau wrote in her blog, "by 6:00 am Berkeley time" she already had an answer from MSRI, and by the end of the day eight other American math institutes were on board. Over the next year, the idea spread by itself to the rest of the world. Rousseau explains the popularity of the idea simply: "The world needed it."

The international climate year, which was called Mathematics of Planet Earth 2013 (MPE2013), eventually grew to include 150 organizations, including math institutes and scientific societies from around the world. Some of the more prominent events that took place during that year were a series of public lectures sponsored by the Simons Foundation, a Mathematics of Planet Earth Day at UNESCO in Paris, and a virtual museum exhibition developed at Oberwolfach in Germany.

One enduring legacy of MPE2013 was the founding of the Mathematics of Planet Earth Center for Doctoral Training (MPE CDT), a collaborative project of Imperial College London and the University of Reading. MPE2013 lent its name, its logo, and a great deal of momentum to a proposal that was still in an embryonic state at the beginning of the year. "By July 2013, when the full proposal was submitted, over 200 people were involved, including scientists, industrial partners, administrators, and company managers," says director Dan Crisan. The British government awarded seed money to the new center in November of 2013, and the MPE CDT welcomed its first class of twelve students (chosen from 158 applicants) in the fall of 2014.

Thanks to new formal academic programs, as well as the thriving informal networks that are now in place, it appears that the days of institutional separation between mathematicians and climate scientists are over.

Sherlock Holmes and John Watson. *(Line drawing by Sidney Paget, 1892. Public Domain.)*

Following in Sherlock Holmes' Bike Tracks

Dana Mackenzie

A MOONLIT NIGHT, A MUDDY DITCH in the English moors, and an enigmatic pair of bike trails were the backdrop for one of Sherlock Holmes' greatest cases—and one of his greatest blunders. In "The Adventure of the Priory School," written by Sir Arthur Conan Doyle in 1905, the fictional detective is summoned to a boarding school in the north of England where a wealthy aristocrat's son has gone missing. Holmes and his sidekick, James Watson, find a pair of fresh bicycle tracks in a field north of the school.

Holmes faces an interesting conundrum: From the tracks alone, how can he figure out which way the bicycle went? Alas, the answer he gives is far from satisfactory:

[Holmes:] "This track, as you perceive, was made by a rider who was going from the direction of the school."

[Watson:] "Or towards it?"

[Holmes:] "No, no, my dear Watson. The more deeply sunk impression is, of course, the hind wheel, upon which the weight rests. You perceive several places where it has passed across and obliterated the more shallow mark of the front one. It was undoubtedly heading away from the school."

"Balderdash!" wrote mathematicians Joseph Konhauser, Dan Velleman, and Stan Wagon in a book called *Which Way Did the Bicycle Go?*, published in 1996. In fact, Holmes fell into two fallacies. The first was his method for identifying the rear wheel track. He could have been duped by a clever villain leaning forward on the bicycle. However, a different foolproof method does exist, which is based on mathematics rather than physics and which does not rely on finding a point where the tracks cross.

The second fallacy is his assumption that identifying the rear wheel path is enough to determine the direction of motion. He never actually explains the rest of his reasoning. But even if we fill in the omission in the only logical way, there is still a problem: "ambidextrous" bike paths can be constructed, pairs of front and rear wheel tracks that can be ridden in either direction. The simplest example is a pair of concentric circles, but there are many more.

Surprisingly, the problem that fooled Holmes has continued to inspire interesting mathematics even 110 years after the story was written, and 20 years after Holmes' solution was busted. Bicycle tracks relate to an extraordinary variety of interesting phenomena in physics and mathematics: from nineteenth-century planimeters to the motion of smoke rings, to floating logs, to a class of mathematical problems called completely integrable systems. "This is a really rich subject," says Sergei Tabachnikov, a differential topologist at Pennsylvania State University. "Whatever direction you look, you find nontrivial questions."

Tabachnikov, along with his co-authors Bob Foote and Mark Levi, recently solved another mystery from the Sherlock Holmes era, which could be called "the mystery of the hatchet planimeter." The hatchet, or Prytz, planimeter (see Figure 1) was a device invented in the late 1800s to measure areas of planar regions. To use the planimeter, you trace the curve with the stylus while letting the hatchet follow along behind as if it were the rear wheel of a bicycle.

In the digital age, planimeters have, sadly, passed into the limbo of obsolete technologies (presumably in a display case right next to the slide rules). But there was a time when they were an obligatory possession for any railway or ship engineer. "The killer application was steam engine calibration," says Foote. "On the side of the steam engine is an indicator that goes around in

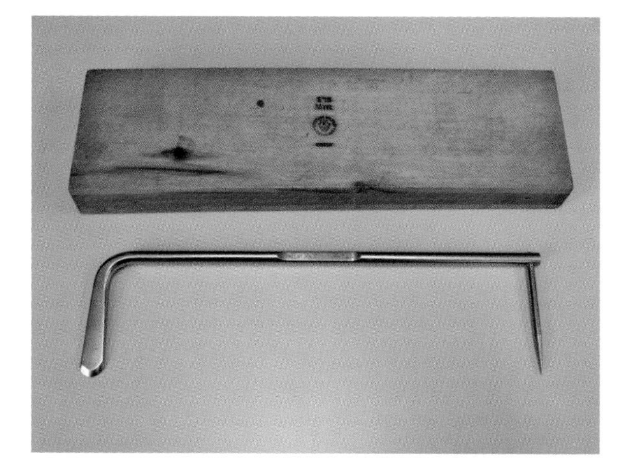

Figure 1. *Planimeters were once a commonly used instrument for measuring areas inside closed curves. (top) A Prytz or hatchet planimeter, an extremely simple design with no moving parts. (bottom) The much more intricate Amsler planimeter, the most popular design in the 19th and early 20th centuries. (Photos courtesy of Robert Foote.)*

a loop, and the area of that loop has to do with the horsepower generated by the engine."

The Prytz planimeter was never the most popular version; another model called the Amsler planimeter dominated the market. One drawback of the Prytz planimeter is that it almost never gives a precise result: it has a systematic error that perplexed its users. Nevertheless, it was much simpler than other planimeters, with no moving parts, so it is understandable that engineers wanted to understand the discrepancy. One such engineer, A.L. Menzin, deduced a correct description of the error in 1908 from empirical observation, but lacked the mathematical skill to prove it. Now (a century too late?) Tabachnikov, Foote and Levi have proved Menzin's conjecture and related it to a mathematical concept called monodromy.

Angular Deflections and Monodromy

The secret behind the planimeter has to do with two very non-obvious properties of bicycles. The first one is this: *the rear wheel of a bicycle never goes faster than the front wheel.* They only go at the same speed when the bicycle is going straight ahead. Most people (including apparently Sherlock Holmes) have never noticed this fact, because the difference in the speeds of the two wheels is normally very slight. It becomes greatest when the front wheel is turned at a 90-degree angle to the frame, forcing the rear wheel to come to a complete stop. Most bicyclists, needless to say, never have to execute such a maneuver.

A remarkable consequence is that the path traveled by the rear wheel, in any finite interval of time, must be shorter than the one traveled by the front wheel. When you ride your bike in a closed loop—say, one lap around a velodrome—the rear wheel does not travel in the same path as the front. It usually travels in a loop *inside* that formed by the front wheel.

Serge Tabachnikov and Mark Levi. *(Photo courtesy of Michael Bialy.)*

Bob Foote. *(Photo courtesy of Robert Foote.)*

Had Sherlock made a more careful mathematical study of bicycle motion, it would have been (to borrow a phrase) *elementary* for him to identify the rear wheel path. If the bicycle were riding a closed circuit, the rear wheel's track would be the interior curve. And even in a one-way trajectory, as in "The Adventure of the Priory School," the front wheel will tend to make larger swerves than the rear wheel. Unless the rider is going in a dead straight line, it will be obvious which is the shorter track, and it has nothing to do with which tire's impression is deeper (because that can be faked). This was the first of Sherlock's two fallacies.

Another counterintuitive property of bicycles is that when the front wheel goes around a closed loop and returns to its original position and orientation, the rear wheel *almost never* returns to its original position. Automobile drivers encounter this situation routinely when parallel parking: they want to turn the car by the largest angle possible within a very small space, usually while traveling less than a full carlength. If you have ever backed into a parking place, driven forward again, and then backed up again at a steeper angle, you have taken advantage of this property.

Likewise, the Prytz planimeter depends on the fact that the device does not return to its original orientation when the front wheel makes a closed loop. The angle of rotation of the planimeter provides (in a manner explained below) a direct estimate of the area enclosed by the loop.

To see how the Prytz planimeter works, it helps to write down the equations that describe its motion (or a bicycle's). A bicycle has a front wheel and a rear wheel, and the position of both of them changes with time. Thus there are four variables in all: the position of the rear wheel at time t, $\gamma(t) = (x(t), y(t))$, and the position of the front wheel at time t, $\Gamma(t) = (X(t), Y(t))$. However, there are two relations between those four variables, one algebraic and one differential. First, because the bike frame's length is fixed, with length L, we can conclude that $(X-x)^2+(Y-y)^2 = L^2$. Alternatively, we can introduce a new variable θ, the angle that the bike frame makes with the x-axis, and write $\Gamma(t) - \gamma(t) = (L\cos\theta, L\sin\theta)$. This reduces the number of variables to three, namely x, y, and θ.

The second relation stems from the one crucial difference between the front and back wheels: the rear wheel cannot turn independently of the bike frame. Thus its direction of motion $(\frac{dx}{dt}, \frac{dy}{dt})$ always points along the oriented direction of the bike frame. Alternatively, the vector $(\frac{dy}{dt}, -\frac{dx}{dt})$ (i.e., the tangent vector rotated by 90 degrees) is perpendicular to the bike frame:

$$(L\cos\theta, L\sin\theta)\cdot\left(\frac{dy}{dt}, -\frac{dx}{dt}\right) = 0.$$

Using differential forms, one can say that $\cos\theta dy - \sin\theta dx = 0$.

Like all planimeters, the hatchet planimeter exploits Green's theorem, which says that the area inside a closed curve is given by integrating the differential form $x\,dy - y\,dx$ around the curve: $A = \frac{1}{2}\oint(x\,dy - y\,dx)$. But unlike in most planimeters, there are two curves in play: with two different areas:

$$A_\gamma = \oint_\gamma (x\,dy - ydx)$$

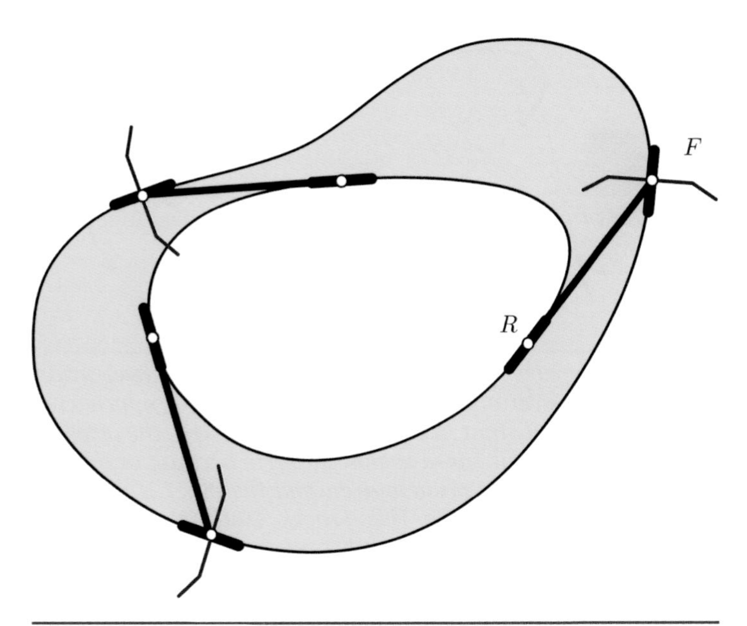

Figure 2. *The two tracks of a bicycle are linked by the fact that the bicycle forms a line segment of constant length, tangent to the rear wheel's track, that ends at the front wheel's track. If both tracks are closed loops, as shown here, the area between them is π times the square of the bike's length. (From "Tractrices, Bicycle Tire Tracks, Hatchet Planimeters, and a 100-year-old Conjecture,"* Amer. Math. Monthly, *vol. 120, No. 3 (March 2013), pp. 199–216. Figure courtesy of Beverly Ruedi.)*

and

$$A_\Gamma = \oint_\Gamma (X\,dY - Y\,dX).$$

The front curve Γ is traced out by the user of the planimeter, and A_Γ is the area that he or she wants to know. However, the planimeter actually measures the area *between* these two curves:

$$A_\Gamma - A_\gamma = \frac{L^2}{2}\int d\theta.$$

The last term ($\int d\theta$) measures how much the planimeter rotates while tracing the loop Γ one time. This angular deflection is easy for the planimeter user to measure.

Two special cases are particularly interesting. First, if the rear wheel should happen to travel in a closed loop as well (see Figure 2), then the angle of rotation is 2π and $A_\Gamma - A_\gamma = \pi L^2$. It's quite striking that the area between the loops in this case is absolutely independent of their shape. A particularly simple case is the one where Γ and γ are circles of radius R and r. Then the above equation makes it easy to see that $R^2 = r^2 + L^2$, which is consistent with the Pythagorean theorem. A second interesting case occurs if A_γ should happen to be zero; then the planimeter measures A_Γ exactly. This is not as far-fetched as it might appear, because the areas in question are signed areas. So if the rear wheel traces a butterfly-shaped region, as in the left part of Figure 3, the areas of the two wings may cancel each other out and give a total area $A_\gamma = 0$.

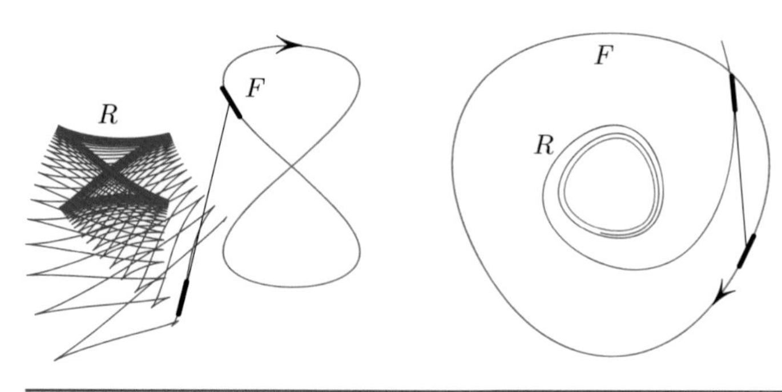

Figure 3. *Hyperbolic monodromy. As the front wheel traces the same loop over and over, the rear wheel track approaches a limit curve. Note that in the example on the left, the oriented area inside the limit curve is close to zero, because one of the "butterfly wings" has positive area and the other is negative. (From "Tractrices, Bicycle Tire Tracks, Hatchet Planimeters, and a 100-year-old Conjecture,"* Amer. Math. Monthly, *vol. 120, No. 3 (March 2013), pp. 199–216. Figure courtesy of Beverly Ruedi.)*

However, in general the area A_y will not be zero and will depend on the starting location of the rear wheel, or hatchet. The question arises, then, how should the planimeter be positioned initially to guarantee that A_y is small? Prytz himself recommended placing the hatchet at or near the centroid (center of mass) of the region whose area is to be found. In fact, it can be proved that the percentage error when starting at this point is bounded by the inverse square of the length L.

Tabachnikov, Foote and Levi looked at the problem from a more modern perspective. They were more interested in the *dynamics* of the bicycle or planimeter than in measuring areas. Each time you ride a bike of length L around a fixed loop so that the front wheel comes back to its starting point, P, the rear wheel will come back to a different point Q. However, Q will always lie on a circle of radius L centered at P. Thus if you simply take a photograph of the bicycle each time it gets back to P, you can reduce the dynamics of the motion to a single function that accepts as input the initial orientation of the bicycle, θ_0, and produces as output the final orientation, θ_1. If you could understand the nature of this function, $\theta_1 = f(\theta_0)$, then you could predict the motion of the real wheel for all eternity without solving a difficult differential equation. Note that the function f depends on the path. To find the orientation of the bike after 1000 laps around the same path, you would just iterate the function f a thousand times. This function f is called the *monodromy* of the dynamical system.

In 1998, Foote proved that the monodromy of a bicycle has a particularly simple form, called a Mobius transformation. This is an extraordinary simplification for two reasons. First, it reduces a complicated dynamical problem to a simple algebraic computation. You can compute an iterated Mobius transformation simply by multiplying matrices. Second, although the space of all curves is infinite-dimensional, the space of Mobius transformations is three-dimensional. That means the monodromy, which superficially seems to require an infinite amount of information about the curve Γ, can actually be described with just

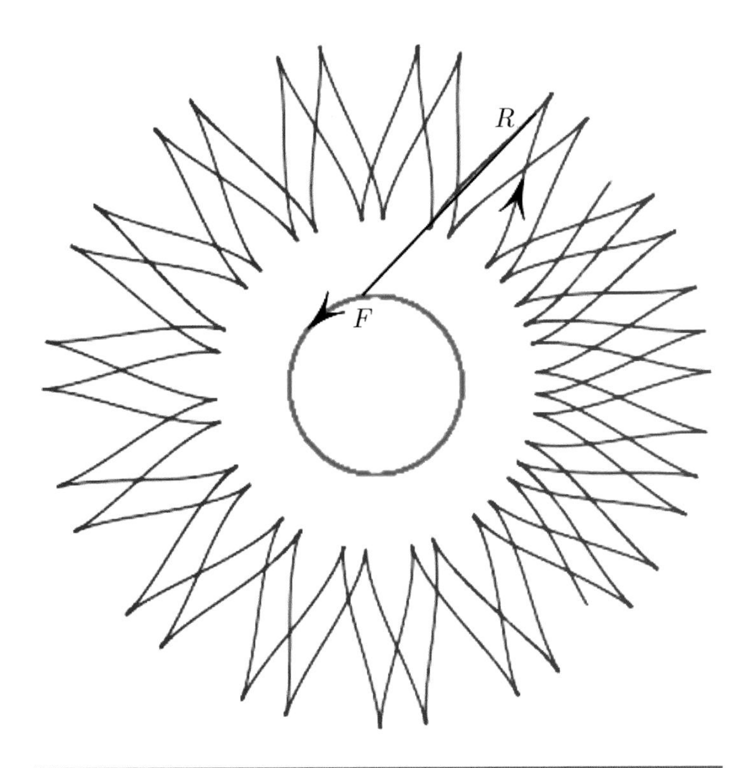

Figure 4. *Elliptic monodromy. Each time the front wheel traces the inner ellipse once, the rear wheel executes one zigzag, traveling first forward and then backward. The rear wheel's path does not "bunch up" as in Figure 3 because there is no limit cycle. (From "Tractrices, Bicycle Tire Tracks, Hatchet Planimeters, and a 100-year-old Conjecture,"* Amer. Math. Monthly, *vol. 120, No. 3 (March 2013), pp. 199–216. Figure courtesy of Beverly Ruedi.)*

three numbers. Later, Tabachnikov and Levi generalized Foote's result to three dimensions. (Yes, it's possible mathematically to ride a bicycle through the air, although nobody outside *The Wizard of Oz* has ever been seen to do it.)

Knowing that the monodromy has this particular form also helped them understand the error term better. In 1908, Menzin noticed that when you traverse the same loop over and over with a planimeter, two distinctly different behaviors can occur. When the planimeter is relatively long compared to the loop, the rear stylus travels around the front wheel's path in an orderly fashion, as shown in Figure 4. This repeated backing up and moving forward is reminiscent of parallel parking. Note that the trajectory does not "bunch up" at any point.

On the other hand, when the planimeter is relatively short compared to the loop, the stylus's trajectory approaches a limiting curve, indicated by "R" in Figure 3. If you start with the stylus off the curve R, it tends to approach the curve. If you started with the stylus exactly on the limiting curve, it would stay on the curve forever, and return again and again to the same starting point.

Thus, the limit curve represents the set of initial points that come back to the same point after one iteration of f: that is, $f(\theta_0) = \theta_0$. Such a point is called a *fixed point* of f. What Menzin

In the context of the original problem, hyperbolic monodromy is bad....However, from the point of view of a modern mathematician, there is nothing bad about hyperbolic monodromy.

observed was a change in the dynamics of the system, from one without a fixed point (this is called elliptic monodromy) to one with a fixed point (hyperbolic monodromy). Furthermore, he wrote, "From purely empirical observations, it seems that this effect [i.e., a limit curve] can be obtained so long as the length of arm does not exceed the radius of a circle of area equal to the area of the base curve." In other words, if $A_\Gamma > \pi L^2$ then the monodromy is hyperbolic. This is the conjecture that Tabachnikov, Levi and Foote have now proven.

In the context of the original problem, hyperbolic monodromy is bad. It means that there are certain starting points that give no angular deflection, and therefore the planimeter reads 0 after one trip around the curve. From this point of view, Tabachnikov's theorem can be seen as a criterion for when the length of the planimeter, L, is too small for the planimeter to work (or equivalently, the area A_Γ is too large for it to measure).

However, from the point of view of a modern mathematician, there is nothing bad about hyperbolic monodromy. The transition from one kind of monodromy to another is exactly what makes the bicycle equation so interesting. And it is miraculous that despite the infinite diversity of closed bicycle paths, one simple inequality can tell you if the monodromy is hyperbolic.

Foote, who dug up Menzin's century-old paper while doing a systematic search for forgotten old articles about planimeters, says that he believes Menzin was motivated more by curiosity about the limit curves than by any actual application. "I think that was his mathematical side showing," says Foote.

Reversing the Direction of Motion: The Bicycle Transformation

Now let's come back to Sherlock Holmes' story. As noted above, there were two fallacies in his treatment of the problem, "Which way did the bicycle go?" and we have only discussed one of them. The second fallacy leads to another interesting feature of the bicycle-tracks problem, whose ramifications are still being explored.

Remember that Holmes' first step was to identify which bicycle track was that of the rear wheel. Although we may criticize the method by which he did this, we must also commend his intuition that this would be an essential piece of evidence. However, it is only a starting point. By itself, it does not tell you which way the bicycle was traveling.

Perhaps Holmes thought the next step was obvious. If you know which track was made by the rear wheel, and you know the length L of the bike, then at any instant there are only two places where the front wheel could have been. If P is the point where the rear wheel is, you draw a tangent line to the rear wheel's track at that point. Then you move a distance L along that tangent line, both forward and back, obtaining two new points Q and Q'. One of those two points must be the place where the front wheel was when the rear wheel was at P. So it is simply a matter of observing which of these two points, Q or Q', actually lies on the observed front-wheel track. If it is the one that lies toward the priory school, then the bicycle was traveling toward the school. Otherwise, the bike was traveling away.

Even Holmes could not imagine a third possibility: *both* of those points might lie on the front-wheel track! In other words, the tangent line to γ at a point P might intersect Γ in two points, Q and Q', each of them the same distance L from P. What rotten luck! But if that had happened, perhaps Holmes would have simply chosen a different base point P, and tried again. Surely he couldn't be unlucky again?

Indeed he could. One simple case was mentioned earlier—the case when both front and rear wheels make a circle. But amazingly, there are many other examples, such as the one shown in Figure 5, next page. The front wheel travels along the black path and the rear wheel along the red. If your bicycle traveled along these paths, neither Sherlock Holmes nor anybody else could ever deduce which way you had ridden. The blue lines show the position of the bike if it is oriented in one way, and the green lines show its position if it is oriented in the other.

However, Sherlock Holmes' sorrow is a mathematician's joy, because unexpected phenomena are always a fertile source of new problems. For example, think of the black curve as the cross section of a log, and suppose that the log has just the right density so that when you throw it in the water, it floats with the water line along one of the blue/green lines in the figure. Then if you rotate it by any number of degrees, it will float with the water line on one of the other blue/green lines! This is surprising because most logs or indeed most floating objects (such as boats) have only a finite number of stable floating positions. But if you can find reversible bike paths, you can also find logs that float in any position.

However, Sherlock Holmes' sorrow is a mathematician's joy, because unexpected phenomena are always a fertile source of new problems.

The latter problem has a long history, and it has yet to be completely solved. It was first proposed in the 1930s, in a legendary circle of mathematicians who used to meet for conversation and problem-solving at the Scottish Café in Lviv (then Poland, now Ukraine). Many examples of floating-in-any-position logs are known, but there is not even any criterion known to determine which densities allow such a log to exist. (It is known that floating-in-any-position logs exist with density $1/2$, but not with density $1/3$.)

The ambidextrous curves, which can be ridden in either direction, are a special case of what Tabachnikov calls the "bicycle transformation." Given any closed front-wheel curve, and its associated back-wheel curve, you can define the "reversed" front-wheel curve to be the curve obtained by riding the bike in the opposite direction along the same back-wheel curve. If the back-wheel curve is ambidextrous, then the front-wheel curve and reversed front-wheel curve are identical, as we have just explained.

But what's surprising is that even if the front-wheel curve is *different* from the reversed front-wheel curve, they nevertheless have an unbelievably long list of things in common. The reversed front-wheel curve has the same length, encloses the same area, and has the same squared curvature integral. In fact, there are infinitely many geometric quantities that the two curves have in common. They're like two people with the same name, address, and social security number.... How do you tell them apart?

As surprising as it is, this phenomenon is not unknown. The same thing happens in the *filament equation*—a differential equation that describes the motion of an idealized smoke ring. A smoke ring, too, does not change its length or its total squared

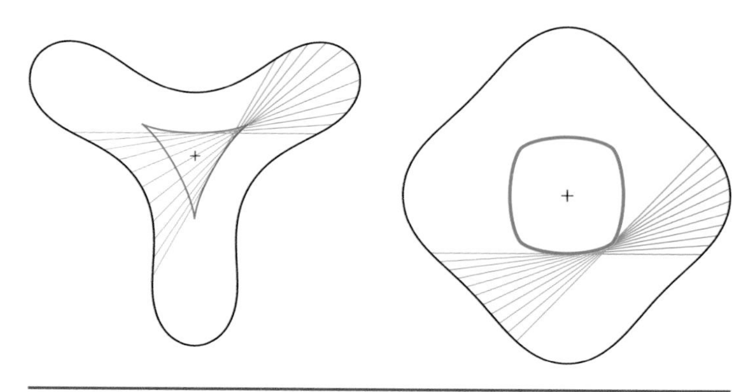

Figure 5. *Two "ambidextrous" bicycle tracks. It is impossible to tell whether the bicycle went clockwise or counterclockwise, because both the blue segments and the green segments have the same length. (Figure courtesy of Franz Wegner.)*

curvature or a variety of other measures so long as it remains intact. Thus the bicycle transformation, in either two or three dimensions, is like a discrete version of the filament equation.

Smoke rings not only retain many of their geometric properties as they evolve, they are also amazingly sturdy. Videos can be found on YouTube of dolphins playing with bubble rings as if they were toys—and the bubbles do not fall apart. Smoke rings, too, can play leapfrog with one another without falling apart. This relates to the fact that the filament equation is a *completely integrable system.*

A classic example of a completely integrable system is the gravitational two-body problem, which was solved by Isaac Newton. A planet orbiting a sun always does so in an ellipse. These systems represent what mathematicians expected when they first started writing down differential equations to describe dynamical systems in the natural world. They expected to be able to find an explicit solution that describes the system at any later time.

However, history had other plans. When mathematicians introduced a third body into the picture (i.e., they studied a solar system consisting of a sun and two planets) it was already frustratingly difficult to calculate explicit solutions by integration. Only in specific cases could they make any progress—for example, the case of a sun, a planet, and a small moon; or the earth, the moon, and a small spacecraft. In this "restricted three-body problem" the third mass is negligible. In 1878, George William Hill showed that small perturbations of a periodic orbit led to a differential equation

$$\frac{d^2x}{dt^2} + p(t)x(t) = 0,$$

where $p(t)$ is a periodic potential function. (In other words, after each orbit the planet returns to a position with the same gravitational potential.) Equations of the above type are still called Hill's equations, even though they have dozens of other applications in physics, science, and engineering. Levi has shown that the motion of a bicycle, given its front wheel trajectory, reduces to Hill's equation, with the periodic potential function being the curvature of the track.

For many years, Hill's equations were consigned to the attic because of a monumental discovery made only a decade later. Henri Poincaré proved that the general three-body problem is *not* integrable. In fact, sufficiently complicated dynamical systems, like those in the real world, are almost never integrable. Real-world systems, such as the weather or the stock market or planetary motions, normally display some form of chaos. But integrable systems can never be chaotic. Like the two-body problem or the restricted three-body problem, they contain enough conserved quantities (quantities like energy and angular momentum that do not change over time) to keep the motion predictable. In the excitement that attended the discovery of chaos in the 1960s and 1970s, equations like the filament equation or Hill's equation, which were the antithesis of chaos, understandably looked like museum pieces.

However, everything that was old becomes new again, and the same has turned out to be true of completely integrable systems.

However, everything that was old becomes new again, and the same has turned out to be true of completely integrable systems. In the 1960s mathematicians discovered that some such equations have unexpectedly resilient solutions, called solitons, which are very much like the nearly indestructible smoke rings and bubble rings. This is true in spite of the fact that the equations of motion are very fragile; even a slight modification makes them no longer integrable.

For mathematicians the solitons were a complete surprise, as was the fact that many real- world systems are completely integrable in spite of the fact that you would expect such systems to be exceedingly rare. This combination has continued to make them a hot topic for twenty-first century mathematics. "The integrability is often related to symmetries," says Boris Khesin, an expert in integrable systems at the University of Toronto. These symmetries are typically expressed as conserved quantities, such as energy. A completely integrable system, such as the filament equation, has a whole panoply of conserved quantities. A "nearby" real-world system may not have so many conserved quantities and yet it inherits the predictability of the completely integrable system. "Integrability is such a strong property that it is not very selfish," Khesin says. "It shares its properties with nearby systems."

But completely integrable systems are still a bit mysterious. When the system is infinite-dimensional, there is not even any agreement on how to define them. Are they systems with lots of conserved quantities, like the filament equation? Or systems whose solutions can be found by integration?

This is why the bicycle transformation may turn out to be important. It acts like an integrable system but it isn't a continuous process. It takes one curve to another, and then another, in a series of discrete steps. It is a unique creature, and possibly a simpler version of a completely integrable system. "The geometry of bicycle tracks may shed new light on some aspects of the theory of Hill's equation," Tabachnikov says. "We hope to develop this connection in the near future."

Meanwhile, if you want to know how Sherlock's story ended.... Well, Sherlock was right (of course). The bicycle *was* heading away from the school, and at the end of its trail he and Watson found a dead body. But the missing boy turned out to be safe, the "victim" of a staged abduction that went horribly wrong. For more, you'll just have to read the story yourself!

Case Closed. *Johannes Kepler, rendered by Robert Bosch at Oberlin College, using an algorithm based on the Traveling Salesman Problem and Bezier curves. The algorithm produces a single closed curve of short length passing through a predetermined set of points. Because it's a simple closed curve, the Jordan Curve Theorem guarantees there are two regions: points inside the curve and points outside. In preparation for project Flyspeck (see text), Thomas Hales gave a formal verification of the Jordan Curve Theorem. (Figure courtesy of Robert Bosch, http://www.dominoartwork.com.)*

Quod Erat Demonstrandum

Barry Cipra

"A poem is never finished; it is only abandoned." So wrote the English poet W.H. Auden, paraphrasing the French poet and polymath Paul Valery. Mathematicians, however, have a somewhat different attitude toward proofs. Ever since Thales took a hard look at geometry (see Figure 1), mathematicians have doggedly pursued the rhyme of reason, even when it takes years to reach a QED. Two recent examples—one still in the works, and one that some seemed willing to abandon—illustrate the point.

In 1998, Thomas Hales, then at the University of Michigan, posted a series of eight papers announcing a computer-assisted proof of a centuries-old problem known as the Kepler Conjecture. Hales had begun work on the project in 1992. It took another five years for the proof to be formally accepted, by the *Annals of Mathematics*. The refereed paper was published in 2005. But for Hales, that was only the beginning.

In 1999, Jim Geelen at the University of Waterloo in Canada, Bert Gerards at the Centrum Wiskunde and Informatica in The Netherlands, and Geoff Whittle at Victoria University of Wellington in New Zealand, began a long slog of their own on a problem called Rota's Conjecture. Fourteen years into the collaboration, they announced that they had all the pieces in place to prove the

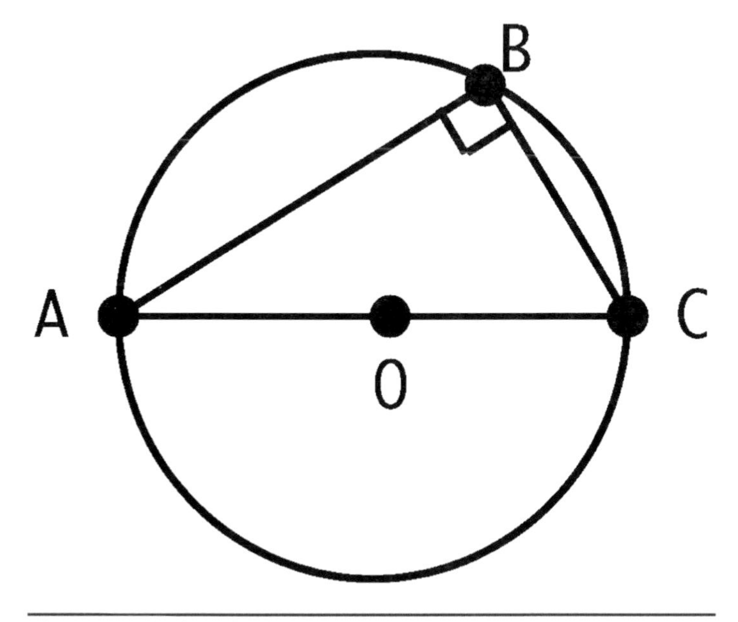

Figure 1. *Thales of Miletus is generally credited with giving the first deductive proof in mathematics. Thales' Theorem says if three points lie on a circle and two of them comprise a diameter, the angle at the third is necessarily a right angle.*

Figure 2. *Stacking oranges densely isn't hard; proving there's no denser way to do it wasn't easy. (Photos courtesy of Bill Casselman.)*

conjecture. Writing up the proof, they estimated, would only take a few more years.

The Kepler Conjecture is a problem that Thales himself could have grasped. It asks for the densest possible packing of nonoverlapping spheres of equal size. In 1611, the astronomer/astrologer Johannes Kepler asserted that one cannot do better than by arranging spheres in layers of hexagonal arrays, with each layer nestling into the triangular hollows of the layer below it—an arrangement familiar to fruit vendors and stackers of cannonballs (see Figure 2). An elementary calculation shows that the fraction of space occupied by such spheres is exactly $\pi/\sqrt{18}$, which comes to approximately 74.048 percent. Kepler himself may have thought the assertion to be self-evident; subsequent generations of geometers did not.

Rota's Conjecture might have been trickier for Thales to comprehend, even though it too has geometric aspects. The conjecture concerns impediments to representing in concrete form—or at least in what passes for concrete form in modern mathematics—a class of abstract objects known as matroids. As the name suggests, a matroid is something like a matrix—and, indeed, matrices provide the so-called concrete representations for some, but not all, matroids. The impediments are known as excluded minors, and if a matroid has one (or more), then it cannot be represented by a matrix (see Box, **"Minor Matters"**). Thales would have had no trouble understanding what a matrix is: It's just a rectangular array of numbers. But he might have been perplexed by the uses they're put to, and also by the types of numbers mathematicians put into their rectangles; in particular, matroid theory makes extensive use of number systems known as finite fields. In 1970, Gian-Carlo Rota, a professor of applied mathematics and philosophy at MIT, conjectured that for each finite field, the list of excluded minors is finite.

Putting a lid on the number of ways that things can go "wrong" is often crucial for understanding the behavior of systems that are otherwise infinite in scope. In 1953, the Hungarian mathematician Laszlo Fejes Tóth observed that the Kepler Conjecture could, in principle, be dispatched in such a way. Tóth reformulated the conjecture in "local" terms, which reduced the problem to a finite collection of questions about finite configurations. This turned the Kepler Conjecture into something a big enough computer could conceivably tackle. At the time, however, computers were big only in physical size; Moore's Law had barely begun to make inroads on what could be practicably computed. But by the 1990s, with *mega-* no longer merely an aspirational prefix for computer power, the time was ripe. Hales reformulated Tóth's reformulation into a suite of calculations that a supercomputer of the day could painstakingly carry out. Three gigabytes of data later, he had all the binary digits to announce a computer-assisted proof.

For Geelen, Gerards, and Whittle, the starting point was major work on minors not for matroids but for a different, more familiar, class of combinatorial objects: graphs. In 1999, Neil Robertson at the Ohio State University and Paul Seymour at Princeton University were in the midst of a huge project to understand excluded minors in graph theory. They had already shown that graphs that can be drawn on a given surface (without

Minor Matters

The concept of a matroid was introduced by the American mathematician Hassler Whitney in 1935, as an abstraction of the notion of independence in linear algebra. A finite set of vectors is said to be independent if none of the vectors in the set is a linear combination of the others. Whitney drew attention to two properties the notion entails: Any subset of an independent set of vectors is also independent, and if two independent sets have different sizes, then the smaller set can be enlarged (while remaining independent) by including a vector from the larger set. Whitney's innovation was to toss away the algebra, keeping only these two abstract properties. A matroid is thus *any* (finite) collection of sets such that the collection contains every subset of every set in it (including the empty set) and if two sets of different sizes are in the collection, then so is a set that consists of the smaller set plus an element from the larger set (not already in the smaller set).

There are many ways to represent matroids. One of them uses their namesake, matrices. If you write down a matrix, the usual linear-algebra notion of independence applied to the column vectors implicitly defines a matroid (see Figure 3). It might seem as if every matroid can be so represented, but that turns out not to be the case; indeed, Whitney's paper gave the first example of a matroid that cannot be represented by a matrix (see Figure 4, page 69). Much of the subsequent theory of matroids has been devoted to understanding what gets in the way.

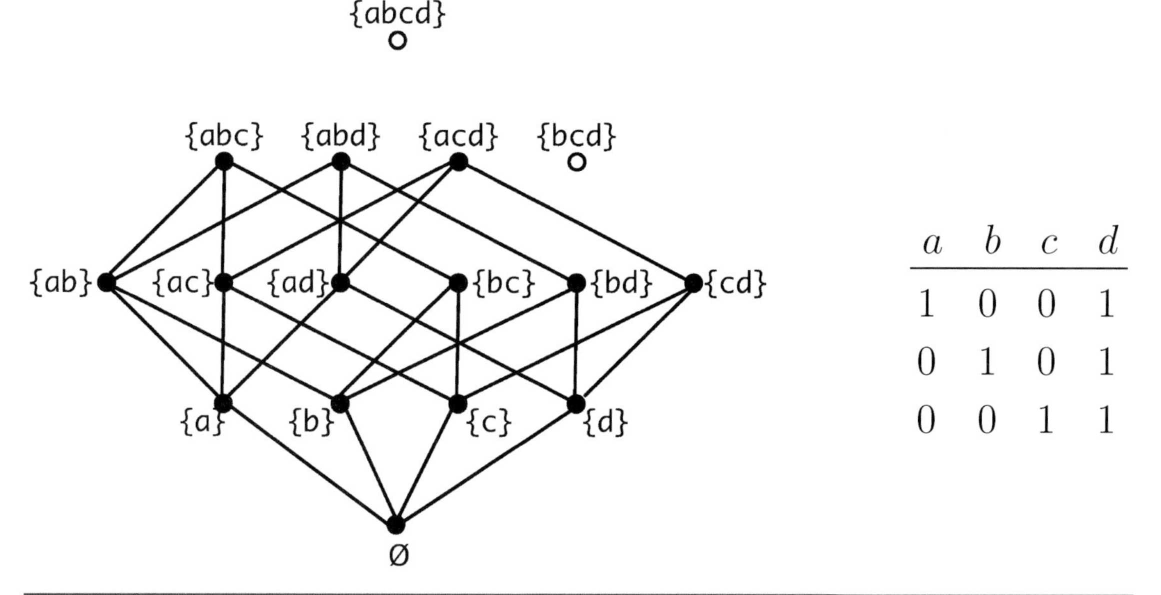

Figure 3. *The independent sets of small matroids are easily listed and/or pictured in full (left), but such lists and pictures do little to reveal their important properties. A matrix representation—if one exists—can be far more informative (right).*

That's where minors come in.

A matroid minor is a subcollection of a matroid obtained by "deletions" and "contractions." A deletion consists of picking an element that belongs to some of the independent sets in the matroid, and then simply deleting all those sets. A contraction consists of picking an element that belongs to some of the independent sets in the matroid, deleting all the *other* sets, and then deleting the chosen element from the ones that remain. When a matroid is represented by a matrix, deletions simply, and obviously, amount to removing columns of the matrix. Less obviously, but equally simply, contractions amount to removing rows as well as columns. Consequently, if a given matroid is representable by a matrix, then so are all its minors. Conversely, if a given matroid is *not* representable by a matrix, then neither is any matroid that contains it as a minor.

This notion of an "excluded" minor has a close parallel in graph theory. Graph minors are formed by a similar process of deletions and contractions (see Figure 5). And much as any minor of any matroid representable by a matrix is also representable by a matrix (with elements from the same field), any minor of any graph that can be drawn on a surface, such as the plane, can also be drawn on the same surface. In particular, Kuratowski's Thereom says that a graph is *planar*—that is, can be drawn on a piece of paper without any edges of the graph crossing one another—if and only if it does not have the graph K_5 or the graph $K_{3,3}$ as a minor (see Figure 6, p. 70).

Neil Robertson and Paul Seymour have pushed the theory of excluded minors far past the plane. Their capstone theorem is that *any* class of graphs that includes all minors of all graphs in it can be characterized by a *finite* list of excluded minors. An identical theorem for matroids is not in the cards; theorists have long known that certain classes of matroids do require an infinite list of excluded minors. But the connections between graphs and matroids run deep, and theorists have succeeded in developing a structure theory for matroid minors analogous to the Robertson-Seymour results on graphs. And that's no minor accomplishment.

any edges crossing) are characterized by a finite set of excluded graph minors, and were on their way to a far-reaching "quasi-well-ordering" theorem about general classes of graphs; their analyses span 23 papers published over three decades. At a conference on graph theory held at the fabled conference center in Oberwolfach, Germany, Robertson and Seymour suggested ways their work could extend to matroid theory. Geelen and colleagues took on the challenge. Over the next decade, meeting off and on for days on end in a seminar room surrounded by whiteboards, the matroid theorists worked at developing a matroid version of Robertson and Seymour's graph minors theory. By 2011, they had much of the machinery in place, but Rota's Conjecture remained out of reach—indeed, their first attempt to prove the conjecture fell short.

Hales, meanwhile, was dissatisfied with one aspect of his published proof: the referees who were tasked with checking the proof ultimately said it was beyond them to certify every

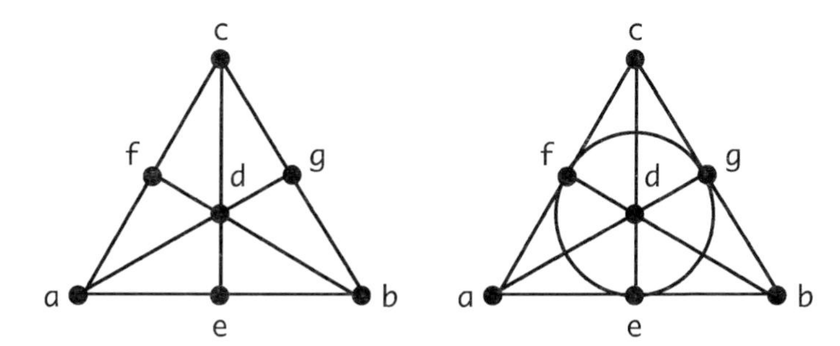

a	b	c	d	e	f	g
1	0	0	1	1	1	0
0	1	0	1	1	0	1
0	0	1	1	0	1	1

Figure 4. *The triangular figure at top left depicts a matroid consisting of all subsets of* $\{a, b, c, d, e, f, g\}$ *with three or fewer elements except for the six subsets* $\{a, b, e\}$, $\{b, g, c\}$, $\{c, f, a\}$, $\{a, d, g\}$, $\{b, d, f\}$, *and* $\{c, d, e\}$, *corresponding to triplets of collinear points. Omitting one additional set (middle), namely* $\{e, f, g\}$, *produces another matroid, known as the Fano plane. Both matroids are represented by the matrix on the right, but over different kinds of fields: The matrix represents the Fano plane only over fields for which* $1 + 1 = 0$. *Can the Fano plane be represented by some other matrix if* $1 + 1 \neq 0$? *In his paper introducing matroids, Hassler Whitney showed that it cannot.*

detail. In response, Hales, who is now at the University of Pittsburgh, embarked on an effort to fill that gap by producing a *formal* proof of the Kepler Conjecture, to be carried out by a computer program that would literally check every single step of his proof. Playing off the initials from "Formal Proof of Kepler," he called the project Flyspeck. The program he picked for the job was HOL Light, developed by John Harrison, an automated reasoning expert at Intel Corporation in Hillsboro, Oregon. The acronym HOL stands for higher-order logic, the *sine qua pauca* of formal mathematics. Harrison's program enables users to restructure existing mathematical proofs in a way that the computer can verify that everything is correct. HOL Light has been applied to a variety of famous mathematical theorems, ranging from the Fundamental Theorem of Algebra to the celebrated Prime Number Theorem. In 2007, as a warm-up exercise, Hales had HOL Light give the thumbs up to the Jordan Curve Theorem. Then, with a team of assistants, he went after Kepler.

The matroid troika, with assistance from postdocs Tony Huynh and Stefan van Zwam, spent much of 2012 developing new machinery specific to Rota's Conjecture. The new machinery hinged on a technical notion for matroids known as connectivity. Roughly speaking, the researchers first showed that the excluded minors in Rota's Conjecture are necessarily highly connected. They then showed that for any given degree

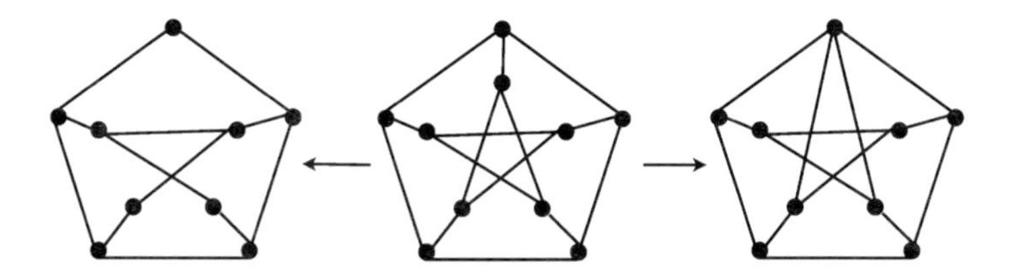

Figure 5. *Minors of a graph (middle) are formed by deleting vertices and edges (left) or by contracting one vertex to another along an edge (right).*

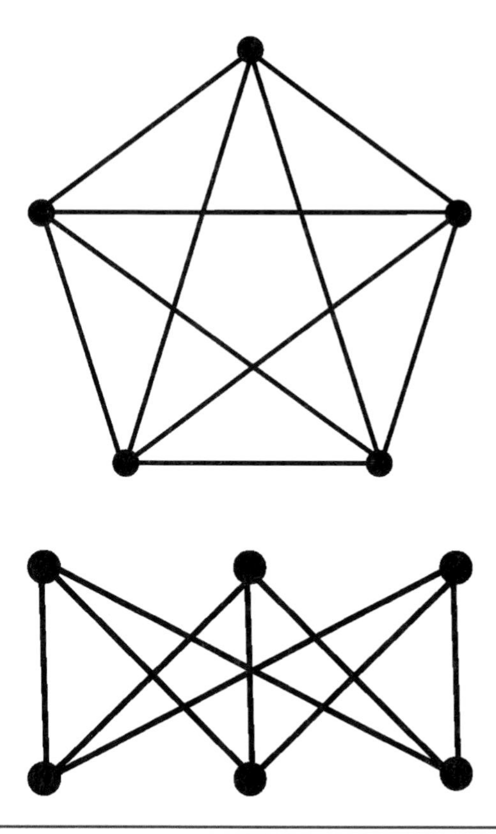

Figure 6. *If a graph contains either K_5 (top) or $K_{3,3}$ (bottom) as a minor, then it cannot be drawn in the plane (without at least two edges crossing). Kuratowski's Theorem says that these two graphs are the only excluded minors for the class of planar graphs.*

of connectivity, the number of excluded minors is finite. The final piece of the puzzle was to put a limit on just how highly connected an excluded minor can be. They announced the proof in 2013—and are now writing it up.

When he first embarked on Flyspeck, Hales estimated the project would take twenty man-years to complete. He was not far off: Working with a large team of collaborators including chief programmer Alexey Solovyev, now at the University of Utah, and chief formalizer Hoang Le Truong at the Institute of Mathematics in Hanoi, Hales had HOL Light ready to test his proof of Kepler's Conjecture in less than a calendar decade. The formal verification ran successfully in 2014. Hales and 21 coauthors have posted a paper describing the proof and explaining the features of HOL Light, along with another verification program called Isabelle, that make the conclusions they reach trustworthy. In particular, the "kernel" of HOL Light, which contains only a few hundred lines of code (next to nothing compared to most operating systems), has been thoroughly vetted; it has even verified its own correctness. "A formal proof in the HOL Light system is more reliable by orders of magnitude than proofs published through the traditional process of peer review," Hales's group writes.

That's not to say there is no longer a role for humans. An HOL Light verification must still be "audited" to make sure, for example, that the statement the program is asked to verify is an exact formulation of the theorem it represents, and that no extraneous axioms have been slipped in. There is also decidedly a role—for now, at least—for humans to compose the "blueprints" for formal proofs: someone has to tell the computer what to verify, in enough detail that the computer can fill in the rest of the steps.

Verification enthusiasts believe that HOL Light and programs like it will become part of mathematicians' toolkit. They are working toward a critical mass of theorems that have been formally checked. Indeed, while Hales was working on Flyspeck, Georges Gonthier at Microsoft Research Cambridge led teams that used a verification program called Coq, developed at INRIA, The French Institute for Research in Computer Science (*Informatique*) and Automation, to check proofs of two deep theorems: the famous Four Color Theorem and a result in group theory known as the Feit-Thompson Theorem. Coq is also at the computational heart of an ambitious project by Vladimir Voevodsky at the Institute for Advance Study in Princeton, New Jersey, to provide what he calls "univalent foundations" for all of mathematics. Voevodsky's project is based on homotopy theory, which originated as a way of understanding topological equivalences (such as why a donut and the coffee cup you dunk it in are topologically the "same") but has become a subject in its own right.

The Coq proof of the Four Color Theorem has special significance—and could be meaningful for the future of matroid theory. The theorem, which colloquially says that every map can be colored with at most four colors, is actually a statement about planar graphs, and it was one of the drivers in Robertson and Seymour's early work on graph minors. It was first proved in 1976, by Kenneth Appel and Wolfgang Haken at the University of Illinois, who developed a theory that reduced the rest of the proof to ruling out a huge (at the time) number of possible counterexamples, which they then wrote a computer program to do. In 1996, Robertson and Seymour, together with Daniel Sanders at Ohio State and Robin Thomas at the Georgia Institute of Technology, came up with a new proof that was far simpler than Appel and Haken's original. However, it still required extensive computation. It was this streamlined but still computer-dependent proof that Gonthier's group massaged into a form that could be formally verified. If programs like Coq and HOL Light can handle intricate results like Kepler's Conjecture and the Four Color Theorem, can matroid minors and Rota's Conjecture be far behind?

To paraphrase another saying of the poet Paul Valery, mathematics is simply the aggregate of all algorithms that are always successful; all the rest is literature.

Gian-Carlo Rota, circa 1970. *(Photo by Konrad Jacobs and courtesy of Archives of the Mathematisches Forschungsinstitut Oberwolfach.)*

Johannes Kepler.

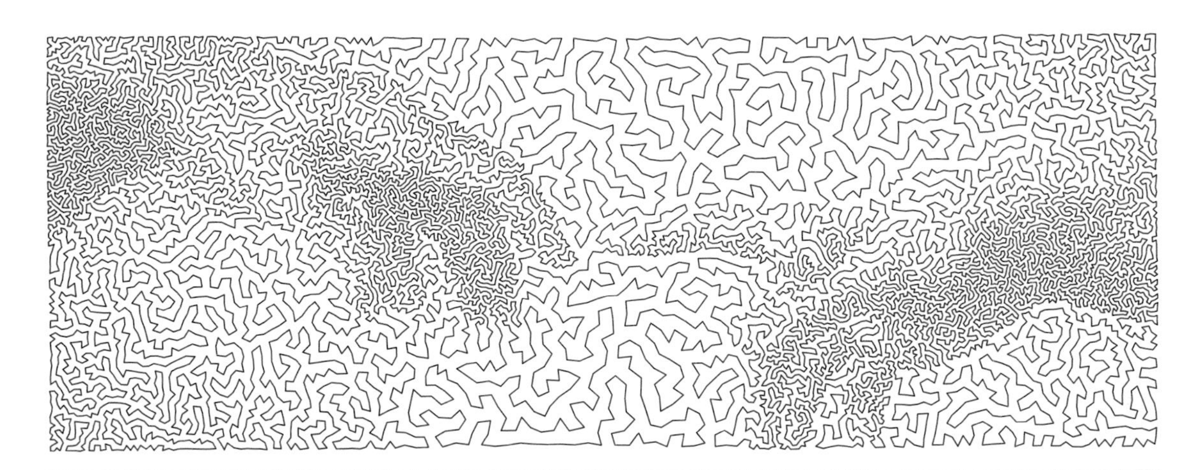

Traveling Salesman Problem. *If each corner in the above path represents a city on a map, then this path is the shortest closed curve joining all the cities. Thus it solves the so-called "Traveling Salesman Problem" (TSP) for this particular map. Much less is known about how to solve the Asymmetric Traveling Salesman Problem (ATSP), in which the measure of "distance" depends on the direction traveled (e.g., some roads are congested or one-way). Thanks to the solution of the Kadison-Singer Problem, mathematicians have now come much closer to an algorithm that would efficiently compute *approximate* solutions to the ATSP. (Figure courtesy of Robert Bosch.)*

The Kadison-Singer Problem: A Fine Balance

Dana Mackenzie

IT'S A FAMILIAR SITUATION TO ANY PARENTS who have two or more children: You want to give toys to each child so that each one gets roughly the same value. But the presents have different values, and you can't apply the Solomonic solution of chopping a toy in half. How close can you come to satisfying each of them?

Now let's make the problem a little bit harder. Suppose you are Santa Claus and you have to distribute toys to *millions* of families. Each family has two children, each child having unique likes and dislikes. You have an identical bag of toys to bring to each family. But you don't have very much time at each stop, so you would like to come up with a *single* way of partitioning the toys into two subsets, that will make *every* child feel fairly treated.

Sounds impossible, right? But Santa is used to doing things that seem impossible, and so are mathematicians. Recently, three mathematicians solved a problem that is reminiscent of Santa's dilemma, called the Kadison-Singer problem. This problem likewise asks for a way to partition a finite set of objects into two or more subsets in a way that (approximately) balances the "value" of each subset, even when the "value" is measured in many different ways.

"It was really by accident that we learned about the Kadison-Singer problem," says Dan Spielman of Yale University, one member of the team that solved it. (All three were originally based at Yale.) "It started when Nikhil Srivastava and I wrote a paper with an undergraduate named Josh Batson on sparsification of graphs."

A graph means a network of points (or "vertices") connected by edges. Think, for example, of the Internet, with the vertices being IP servers and the edges being the fibers that connect them. A "sparsification" is a subgraph that has the same vertices but not as many connections. For example, you might assign roughly half the edges to one subgraph and the other half to the complementary subgraph. The challenge is to make these two sparser subnetworks "behave" roughly the same way as the original; for example, messages should go from one place to another in about as many jumps. Then you have two redundant networks where you once had only one—a desirable feature if some of the connections should happen to fail.

Spielman explained the problem to a friend of his, Gil Kalai of Hebrew University, who commented that it reminded him of other problems that have to do with partitioning discrete sets into balanced subsets. For example, there was the Paving Conjecture, which has to do with partitioning the rows and columns of a matrix; the Weaver Conjecture, which has to do

"…. It was a fluke that every time I entered a new area of research, this was equivalent to their most famous unsolved problem!"

with partitioning a set of vectors; and the Feichtinger Conjecture, about partitioning signals into wavelets. In fact there are so many problems of this type that Pete Casazza of the University of Missouri had compiled a list, in 2006, of more than a dozen conjectures in different branches of math, all of which boiled down to one problem: the Kadison-Singer problem.

"I was not looking for equivalent versions of the Kadison-Singer problem," Casazza says. "I was trying to solve it, and I started moving into different areas of research, hoping that one of them had deep enough results to handle the problem. It was a fluke that every time I entered a new area of research, this was equivalent to their most famous unsolved problem!" Or perhaps not such a fluke. A hard problem is hard for a reason; if it stumps one set of mathematicians, it will probably stump others.

The original version of the problem, posed by Richard Kadison and Isadore Singer in 1959, does not look even remotely like a partitioning or balancing problem. It came up in the context of infinite-dimensional algebras, called C^*-algebras, which are used in theoretical physics. Kadison and Singer realized that an assumption made by the famous physicist Paul A. M. Dirac was mathematically flawed. This is not uncommon; theoretical physicists often take mathematical shortcuts. Dirac, however, was an exceptionally careful physicist, and so he was mistaken in an unusually interesting way. His assumption was easily seen to be valid for finite-dimensional algebras. Kadison and Singer proved that he was mistaken when the algebras have an uncountable infinity of dimensions ("continuous" algebras).

But they were not sure what happens in the intermediate case, where the dimensions are discrete but infinite. Lacking any evidence one way or the other, except for the fact that they had just proven Dirac was wrong in the continuous case, they cautiously wrote: "We incline to the view" that Dirac would be wrong in the discrete case as well. (As it turned out, their hunch was wrong, and Dirac was right.)

Even today, Kadison has a diffident attitude toward the problem that bears his name. When lecturing about it, he emphasizes that he and Singer considered the important part of the 1959 paper to be their *theorem*—i.e., the fact that Dirac was wrong in the continuous case—rather than the problem they left unsolved. But the history of mathematics is fickle, and sometimes even an offhand remark by one mathematician can turn into a holy grail for future generations. "We didn't think the other problem was that important," says Kadison. "We were wrong, but at least we were clever enough to ask the question."

From Paving to Frames

Over the years from 1959 to 2013, the Kadison-Singer problem went through a number of transformations. In 1979, Joel Anderson of Pennsylvania State University posed a new version called the "Paving Conjecture," first an infinite-dimensional version and then one in finite dimensions. While C^*-algebras remain a somewhat arcane subject, finite-dimensional vector spaces are a part of every mathematician's education. Thus the Paving Conjecture made the Kadison-Singer problem accessible to people in a much broader range of specialties. That includes Spielman, who candidly admits, "I don't understand all the formulations of this problem. I literally understood the fraction of the machinery that I needed."

The finite-dimensional version of the Paving Conjecture has to do with splitting n-by-n matrices into a smaller number, r, of tiles that cover the main diagonal of the matrix. (Hence the idea of "paving." As shown in Figure 1, the tiles form something like a paved walkway.)

Any matrix A has an operator norm, which is the maximum factor by which it stretches any vector. A geometric way to think about it (see Figure 2, next page) is that the linear transformation represented by A will take the unit sphere (a sphere of radius 1, centered at the origin) to an ellipse, perhaps stretching some axes and perhaps shortening others. The length of the longest axis of the ellipse is the operator norm of A (denoted $\|A\|$).

In the Paving Conjecture, the main diagonal of A is assumed to be zero. So if you isolate the i-th row and i-th column of A, you get the operator 0. This "isolation" operation is done by a projection Q_i into the 1-dimensional subspace of R^n spanned by the basis vector e_i. Of course, the operator norm of the 0 matrix is 0. Symbolically, $\|Q_i A Q_i\| = 0$. More generally, if you let V be any k-dimensional coordinate subspace of R^n and let Q_V denote the projection of R^n into that subspace, then $Q_V A Q_V$ will be a k-by-k submatrix of A. These submatrices are the paving stones for the Paving Conjecture.

For reasons having to do with the original Kadison-Singer problem, Anderson conjectured that the norms of the submatrices would become smaller and smaller as the number (r) of paving stones increased. A naive way to look at it would be that piece 1 would contribute $1/r$ of the norm, piece 2 would contribute $1/r$ of the norm, and so on. (This begins to look like a balanced partition problem.) However, operator norms don't

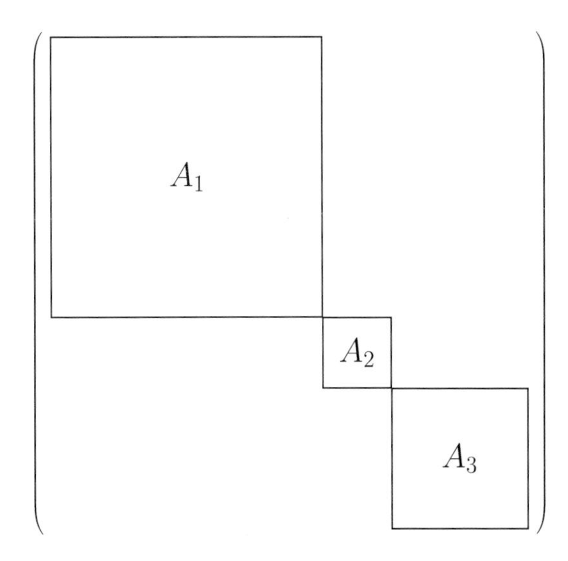

Figure 1. *Paving Conjecture. The object of the conjecture is to split the matrix A into submatrices, here shown as A_1, A_2, and A_3, which have uniformly smaller mapping norms than the complete matrix A. (Figure courtesy of Adam Marcus.)*

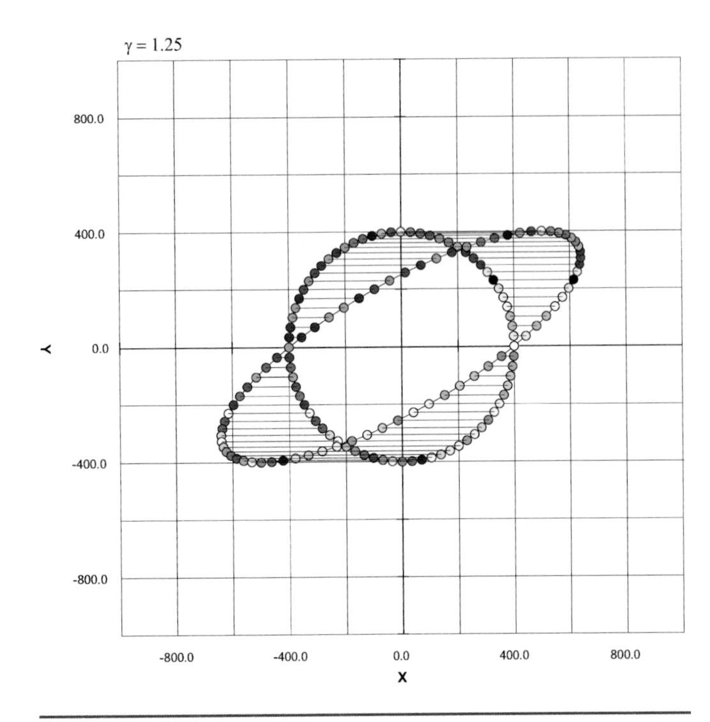

Figure 2. *Mapping norm (or operator norm) of a linear transformation. In this case, a shear transformation pushes each colored dot on the unit circle horizontally to a correspondingly colored dot on the ellipse. The dot that ends up farthest from the origin (at the upper left of the ellipse) is light blue. The mapping norm of the transformation is the distance to that light blue point, or equivalently, the length of the major axis of the ellipse. Mapping norms in high-dimensional spaces are relatively difficult to calculate, and the Paving Conjecture can be interpreted as an attempt to decompose them uniformly into simpler parts. (Figure from Visual Strain 2014, a C# program written by B. Denton and G. H. Girty, Department of Geological Sciences, San Diego, CA 92812.)*

actually add in this way, so we'll simply ask each piece to contribute less than 100 percent of the norm.

Of course, if we simply took $r = n$ we could pave A with 1-by-1 tiles in which every piece contributed 0 to the norm, and we would be done. However, there is a catch. In the Paving Conjecture, we aren't allowed to change r as n grows. Thus the challenge is to split R^n up into r subspaces, V_1 through V_r, so that for each subspace, $\|Q_{V_i} A Q_{V_i}\| < (1 - \epsilon)\|A\|$. For example, if $\epsilon = 0.1$ we achieve a 10 percent reduction in the operator norm. If $\epsilon = 0.2$ we get a 20 percent reduction, and so on. The bigger a reduction we want the more tiles we will need, but Anderson conjectured that for any fixed ε there will always be an upper bound on the number of tiles required, no matter how large the dimension n might be.

It is this independence of the size of the matrix that makes the problem scale up well to infinite dimensions. The infinite-dimensional property is really a finite-dimensional property in disguise.

Anderson's conjecture was more intelligible than the Kadison-Singer problem, but not necessarily any easier. In fact, there are some concrete versions that are still unresolved. For instance, what if we want to reduce the operator norm of A by 10 percent? Can we always do it with three tiles? No example is known of a matrix that needs more. On the other hand, if you want to reduce the operator norm by 14 percent, there is a known example of a 13-by-13 matrix that requires a minimum of four tiles.

Now that Spielman's team has solved the Kadison-Singer problem, Anderson's Paving Conjecture is known to be true as well. We can be sure that *some* number of tiles is sufficient to reduce the operator norm by 10 percent, no matter what matrix A you start with. But we don't know the number. It may be three, or it may be three million. Spielman's group did prove a quantitative version, but he says, "Our estimate of the number of tiles is very far from what we believe the correct estimate to be."

In 2003, Nik Weaver of Washington University in St. Louis derived yet another alternative version of the Kadison-Singer problem that made the connection to balanced partitions even clearer.

In 2003, Nik Weaver of Washington University in St. Louis derived yet another alternative version of the Kadison-Singer problem that made the connection to balanced partitions even clearer.

Weaver's version supposes that you have what is called a *frame* of vectors in R^n. To understand what this means, it is helpful to start with orthonormal frames, which every linear algebra student learns about. In an n-dimensional space, an orthonormal frame is a set of n vectors, $\{u_1, \ldots, u_n\}$, that form the edges of an n-dimensional cube. Orthonormal frames have nice algebraic properties as well as geometric properties. For instance, it is easy to write any other vector x as a "linear combination" of them:

$$x = (x \cdot u_1)u_1 + (x \cdot u_2)u_2 + \ldots + (x \cdot u_n)u_n.$$

Such a representation is handy if you want to split up, or analyze, a vector into simpler parts. For example, if your vector is a radio signal, you might want to split it up into signals at a finite set of frequencies so that you can hear what is being said on each frequency. For other applications you might want to go the other way: a music synthesizer "builds up" a sound (say, a synthesized violin sound) by adding together pure frequencies. Orthonormal frames make both directions, analysis and synthesis, very easy. By contrast, for non-orthonormal coordinate frames, either the analysis or the synthesis will be more time-consuming.

Most linear algebra textbooks do not, however, mention that there is no reason to stop at frames of n vectors. It is possible to construct sets of m vectors (where m may be much greater than n) that satisfy exactly the same equation as above with n replaced by m. Being a more general version of orthonormal frames, such sets of m vectors are simply called *frames.* The price is that the representation of x is no longer unique.

However, for many applications the non-uniqueness is a small price to pay for *redundancy.* A conventional orthonormal frame, $\{u_1, \ldots, u_n\}$, is very brittle. If one component of the data is missing, then the synthesis step will not work; you will get a corrupted message (or you will not get the vector x back). Users of the Internet are very familiar with this phenomenon, for example when a graphic fails to load properly. On the other

"We worked on it for five years of pretty hard work, but it was a very rewarding problem," Spielman says. "We kept feeling as if we were making progress. It was just as if the problem were designed by one of those people who makes video games. Every six months we would make substantial progress, just enough to keep us working on it."

hand, if you have a highly redundant frame, $\{u_1, \ldots, u_m\}$ with m much greater than n, then the failure of one component to transmit correctly doesn't matter so much. You may be able to reconstruct your message, either exactly or approximately, even without it. This redundancy of frames should sound familiar; it is similar to the ability of a graph sparsification to impersonate a denser network.

Weaver's version of the Kadison-Singer problem simply formalizes one aspect of this redundancy. If we split a message up and transmit it through several channels $u_1, \ldots, u_m$, then the total energy used can be written as the sum of squares of the coefficients of the analyzed vectors:

$$\|x\|^2 = (x \cdot u_1)^2 + \ldots + (x \cdot u_m)^2.$$

This is a familiar property of orthonormal frames that remains true for general frames as well. According to Weaver's Conjecture, any frame (subject to certain mild conditions) can be split up into two subframes that split up the energy more or less evenly. In other words, if one of the subframes is $\{u_{i_i}, \ldots, u_{i_k}\}$, then

$$(x \cdot u_{i_1})^2 + \ldots + (x \cdot u_{i_k})^2 \approx \|x\|^2/2.$$

If this were true only for *one* vector x it wouldn't be too surprising. That would be a case similar to one family that has to divvy up its toys so that each child will be equally happy. (In this comparison, "happiness" is the analogue of "energy.") The astounding thing is that the balanced division of energy holds for *any* vector x. This feat is analogous to Santa making every family happy at the same time.

Of course, just like Anderson, Weaver had only rephrased the problem. In 2003, it was still unknown whether this Santa-like balancing of happiness was feasible. In fact, many people continued to doubt it; just like Kadison, they "inclined to the view" that the Weaver Conjecture and the Paving Conjecture would turn out to be false. "I had never done any work on what would happen if Kadison-Singer were true!" says Casazza. "I thought it was false."

This was about the point where Dan Spielman and his team at Yale University entered the picture. His undergraduate student, Batson, had graduated, but then a new postdoctoral researcher arrived who was highly motivated to solve the Kadison-Singer problem. His name was Adam Marcus.

"Adam took a huge risk," Spielman says. "He came here on a non-tenure-track position. He didn't work on anything else—this was putting all of his eggs into one basket. I was very worried about him, and tried to encourage him to work on other things. But he wouldn't do it, because this was too fun."

In fact, Spielman and Srivastava were also caught up in the fun. "We worked on it for five years of pretty hard work, but it was a very rewarding problem," Spielman says. "We kept feeling as if we were making progress. It was just as if the problem were designed by one of those people who makes video games. Every six months we would make substantial progress, just enough to keep us working on it."

It also helped that Weaver's version of the conjecture was so concrete that they could do lots of computer experiments. These experiments showed when they were and weren't on the right track.

Dan Spielman, Adam Marcus, and Nikhil Srivastava.

A Pinch of Randomness

Now let's look at the precise version of Weaver's Conjecture that was proved by Marcus, Spielman and Srivastava, as explained in Srivastava's blog:

Given a frame of vectors, $u_1, \ldots, u_m \in R^n$, such that each vector has length less than $\sqrt{\alpha}$, there exists a partition $T_1 \cup T_2 = \{1, 2, \ldots, m\}$ satisfying the inequality

$$\left| \sum_{i \in T_j} (x \cdot u_i)^2 - \frac{1}{2} \right| \leq 5\sqrt{\alpha}$$

for every unit vector x. (The constant 5 has since been improved to 3.) Note that the inequality holds for *both* T_1 and T_2; they are dividing the energy nearly half-and-half. The left-hand side is the *discrepancy*, a measure of how much we miss a balanced division. Although we can't get the discrepancy to be zero, the right-hand side tells us that the discrepancy is controlled by the parameter α. In the toy analogy, this would be the maximum value of any toy. The theorem gives away Santa's secret: He should make his bag contain a whole lot of really cheap toys. The cheaper they are, the more effectively he can balance the desires of each child. Or to put it another way, if his bag contains one toy that every child really wants, then the ones who don't get it are going to be unhappy.

This tells us part of Santa's magic, but it doesn't give us the complete picture. How is he going to figure out the two subsets of toys, T_1 and T_2?

Unfortunately, mathematicians are not Santas, and they don't have a very huge bag of tricks. The first and simplest strategy for finding the sets T_1 and T_2 is no strategy at all: just pick them at random. This random strategy had been tried by a number of mathematicians before Spielman, Srivastava and Marcus. It didn't work. The best estimate that these mathematicians had proved was that the discrepancy was less than $C\sqrt{\alpha \log n}$, for some constant C. That might be good enough for Santa but it wasn't good enough for Kadison and Singer,

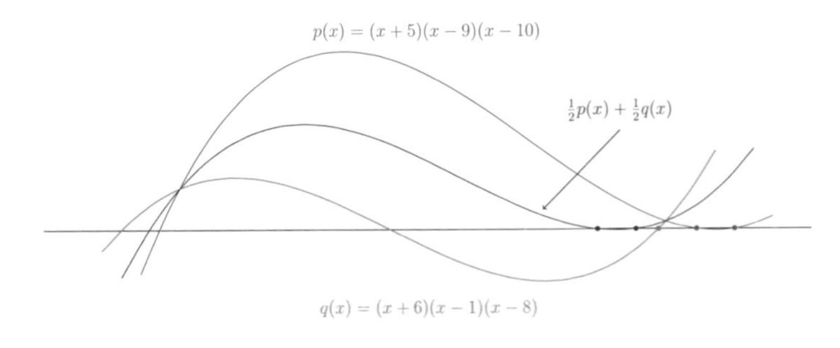

Figure 3a. *Non-interlacing polynomials. The red and green polynomials do not interlace. Remarkably, each one of them is "above average," in the sense that the largest root of each one is greater than the largest root of the average (plotted in blue). This counterintuitive behavior, which one might call the "Lake Wobegon Property," is not possible for interlacing polynomials. (Figure courtesy of Adam Marcus.)*

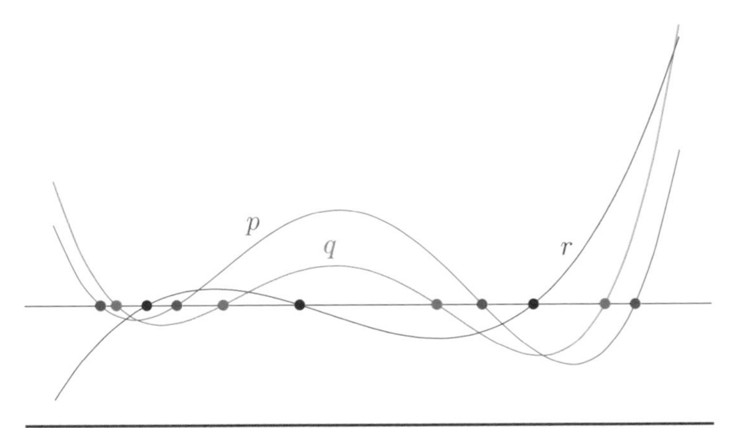

Figure 3b. *Interlacing polynomials. Here, the red polynomial p and the green polynomial q interlace. That is, a third polynomial can be found (r) so that between each pair of consecutive roots of r (blue dots) lie exactly one root of p and one root of q (red and green dots). (Figure courtesy of Adam Marcus.)*

because the estimate needs to be independent of the dimension n.

But while the all-out random approach doesn't work, Spielman's group was able to show that a strategy with just a pinch of randomness does. We can't pick T_1 and T_2 completely at random—because most times we will fail—but we will succeed *some positive percent* of the time. Even if we succeed only one time in a million, that is enough to show that success is possible.

This was not the only innovative idea that Spielman's group came up with. Another very sticky issue was finding the value of x that maximizes the energy function, $\sum_{i\ T_j}(x \cdot u_i)^2$. This turns out to be exactly equivalent—once again—to the problem of finding the operator norm of a matrix A. But Spielman's group came up with several innovative methods to do it.

• First, they realized that they could "build up" the matrix by adding one vector u_i at a time to the subframe T_j. They called these intermediate steps "rank-one updates" to the matrix.

• Second, they realized that merely tracking the operator norm, or the longest axis of the ellipsoid, was not good enough. To understand the behavior of the rank-one updates, they needed to track the lengths of *all* of the axes of the ellipsoid. These lengths are called eigenvalues of A.
• Third, they realized that keeping track of any individual eigenvalue is hard. The eigenvalues are given as roots of a degree-n polynomial, called the *characteristic polynomial* of the matrix, $\chi(A)$. For high n there are no formulas for these roots, and even for low n the formulas are very complicated and non-linear. Instead, it is much easier to follow the evolution of $\chi(A)$ itself as A receives rank-one updates.
• Next, they let the vectors u_i be chosen randomly on a sphere. This meant that the rank-one updates were also random. They let $E[\chi(A)]$ denote the "average" or "expected" characteristic polynomial formed by random rank-one updates to A. This average characteristic polynomial has a beautiful, simple formula, written in terms of a classically known set of polynomials called the Laguerre polynomials.
• Most importantly, they showed that the roots of all of the updated polynomials at various stages of the process "interlace" in a very simple way, illustrated in Figure 3b. This prevents the roots from wandering too far away from the roots of the Laguerre polynomials.
• Finally, they had to go back from average polynomials to individual polynomials. Intuitively, you might expect that one of the randomly generated polynomials must be "below average," in the sense that its greatest root is less than the greatest root of the average polynomial.

The last step, which seems so harmless, was actually the hardest part of the proof. In fact, polynomials inhabit a Lake Wobegon-type world where every kid can be above average. That is, it is possible to find sets of polynomials such that *every single one* has a root larger than the largest root of the average polynomial (see Figure 3a).

However, if the polynomials are interlacing this Lake Wobegon scenario is impossible. At least one polynomial will have its largest root less than the largest root of the average polynomial. To prove this, Spielman's group had to develop a whole theory of interlacing polynomials that didn't exist before. That was one reason that it took them 5 years.

The end result of this argument is that some small fraction of the matrices A formed by this process of random rank-one updates will have "below-average" norms, and this can be used to show that they account for less than 100 percent of the energy (in fact, closer and closer to 50 percent as α gets smaller).

So why did Spielman, Srivastava, and Marcus succeed when so many other people had failed? Casazza suggests an interesting three-part theory. First, they were smart. "I think these guys classify as geniuses," Casazza says. Second, they had incredible perseverance, especially Marcus, who worked on nothing but this problem for four years, risking professional suicide. Third, and most important, "they had the right tools," Casazza says. "Eigenvalues have been studied for 100 years, but we know very little about how the eigenvalues of an operator, a sum of weighted projections, compare to the eigenvalues of a subset."

The Marcus-Spielman-Srivastava theorem could have major consequences for signal processing, where it could speed up the algorithms used to analyze and synthesize signals.

The key was to look at the characteristic polynomials and show they interlaced—"the right tool" for the problem.

The Marcus-Spielman-Srivastava theorem could have major consequences for signal processing, where it could speed up the algorithms used to analyze and synthesize signals. One of the great goals in signal processing is to solve the "cocktail party problem," to take a recording of many people talking and pull out what each person is saying. This appears to require massively redundant frames. Spielman's result shows that these frames might be separable into simpler subframes. However, it is too soon to say what the practical consequences will be, in part because the proof depends so heavily on random choices. "Randomization is a catastrophe for algorithms," Casazza says. Spielman demurs: "I don't care if it's random or deterministic, any more than I care whether it's on a Mac or a Unix." But he agrees that their proof of the existence of balanced partitions cannot yet be turned into a practical algorithm to find the partitions, because it would take too long ("exponential time").

Meanwhile, the theorem has already had significant repercussions in graph theory, where it leads to improved approximations to the solution of the asymmetric Traveling Salesman Problem (see Box, **"Calling All Salesmen"**). It also improves the original theorem that led Spielman to the problem, the Batson-Spielman-Srivastava theorem. Now a graph can be sparsified cleanly, without resorting to subgraphs that contain a fraction of an edge.

Great problems come in many disguises. Is Kadison-Singer a problem about C^*-algebras? Physics? Graphs? Discrepancies? Isotropic sets? Interlacing polynomials? Ultimately, it is all those things, and probably even more that haven't been thought of yet. That is one of the delights of mathematics: One toy can make so many people happy.

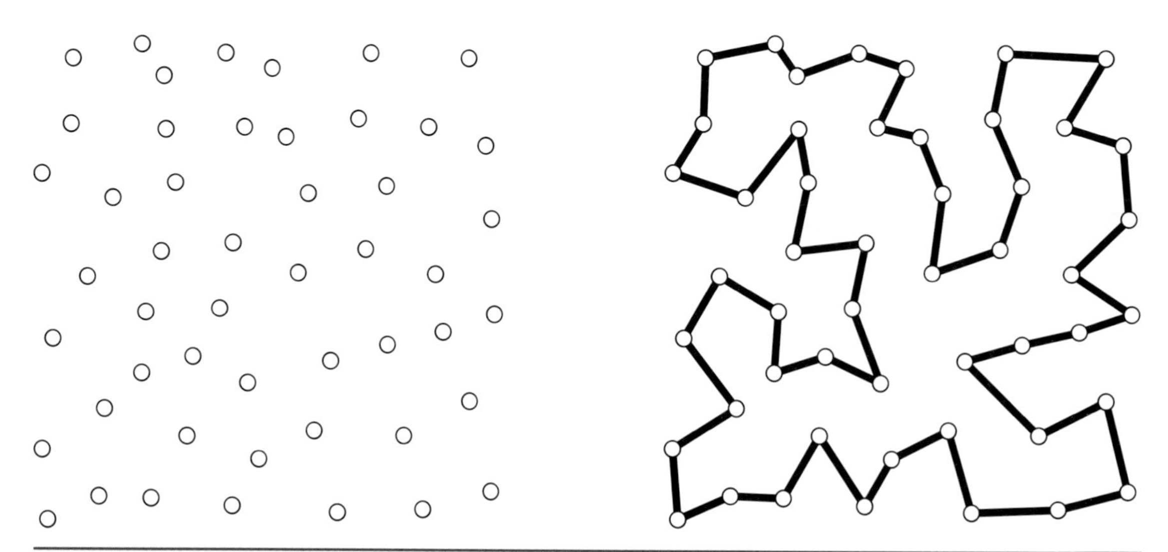

Figure 4a. *A simple instance of the Traveling Salesman Problem. The shortest closed path joining the 50 cities (left) is the black path at right. In this case the distance between cities is assumed to be Euclidean, and it is therefore symmetric (the distance is the same in either direction). (Figure courtesy of Robert Bosch.)*

Calling All Salesmen

If you've been lucky enough to solve one hard problem, often it will provide the key to unlock another. That was the thinking of Shayan Oveis Gharan, a computer scientist at the University of Washington, and Nima Arani of the University of California at Berkeley.

For decades, one of the quintessential "hard problems" in computer science has been the Traveling Salesman Problem (TSP) (see Figure 4a). It's a simple question: given n cities on a map, how can a salesman visit each one of them while traveling the minimum number of miles? The problem can be solved by trial and error, but with even a modest number of cities the number of possible routes becomes so immense that such a solution is impractical. In fact, the TSP has been proven to be an "NP-hard" problem, which means there will probably never be an algorithm that solves it efficiently for any number n of cities.

However, even the Traveling Salesman Problem is easy compared to the *Asymmetric* Traveling Salesman Problem (ATSP) (see Figure 4b, next page). In this version, each road is assigned a cost function, and the cost of traveling north (for example) might be different from the cost of traveling south. An extreme—and instructive—example is the case where all the roads are one-way; in this case, the cost of traveling in the wrong direction is infinite... or, one might say, the death of a salesman.

In 1976, Nicos Christofides of Imperial College London devised an algorithm that is guaranteed to come within 50 percent of the best solution to the ordinary, symmetric Traveling Salesman Problem. For instance, if the best route takes 8000 miles, Christofides' algorithm is guaranteed to find a route that is at most 12,000 miles. However, his algorithm does not work for the ATSP, because it may produce a route that travels the wrong way down a one-way street.

Another approach is called a "relaxation" of the problem, where the salesman is allowed to split himself into fractions and go in several directions at once, subject to certain conditions. Though unrealistic for the real world, such fractional solutions *can* be computed efficiently. Finding them is not an NP-hard problem. Ideally, they would provide a good estimate of the true optimal cost. That is, the "integrality gap"—the ratio between the actual optimal tour (where the salesman stays in one piece) and the best known fractional solution—should stay bounded. Yet even this is not known for the ATSP. Until recently, the best estimate of the integrality gap increased in proportion to the logarithm of the number of cities ($\log n$).

Oveis Gharan and Anari have now used the technique developed by Spielman's group to whittle the integrality gap down to a power of the logarithm *of the logarithm* of n. While it is still not constant (like Christofides' 50 percent), $\log \log n$ grows incredibly slowly as n increases; for practical purposes it is almost indistinguishable from a constant. Nick Harvey of the University of British Columbia calls their work an "amazing result."

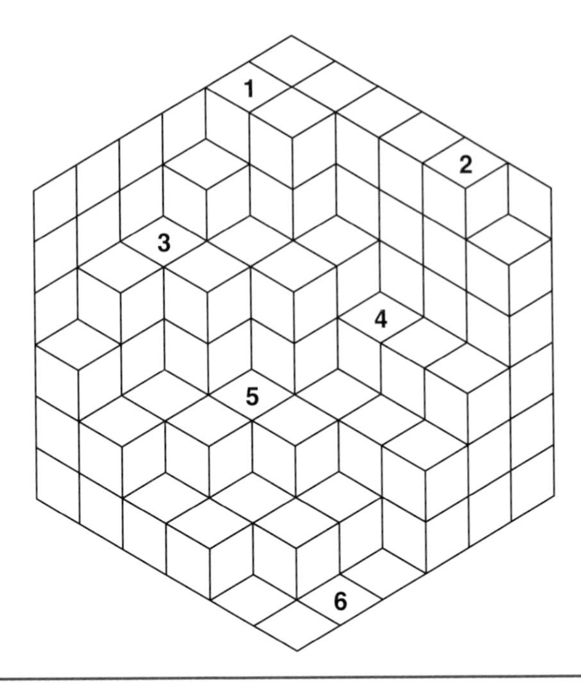

Figure 4b. *Asymmetric Traveling Salesman Problem. A simple instance of the Asymmetric Traveling Salesman Problem (ATSP). The figure should be interpreted as a stack of cubes, with 6 labeled horizontal squares. On any one "move," the salesman can travel to any adjacent horizontal square that is zero to two levels below him. This costs 1 cent. Or he can travel to any adjacent square that is one level above him, but this costs 2 cents. Larger drops or rises on one move are not allowed. Thus the cost of going* $1 \to 3$ *is 2 cents, but the cost of going* $3 \to 1$ *is 4 cents. The minimal-cost loop connecting all six labeled squares is* $1 \to 2 \to 6 \to 5 \to 4 \to 3 \to 1$ *and costs 30 cents. (Figure courtesy of Robert Bosch.)*

To see how the Kadison-Singer problem enters graph theory, let's start with Christofides' algorithm for the TSP. The very first step is to find a minimum spanning tree—a connected subgraph of the highway network that passes through every city and has no loops. (There are several steps after this, but we will skip them for brevity.)

If some routes are one-way, as in the ATSP, it is necessary to match the number of ingoing legs and outgoing legs in each city, so that the salesman doesn't get stuck in Omaha. In the minimal spanning tree, some cities have excess ingoing edges (I) and some have excess outgoing edges (O). The edges of the graph that cross from I to O are called a "cut." Because extra edges have to be added to balance I and O, which worsens the performance of Christofides' algorithm, it is good to start with an *alpha-thin* tree: a tree that contains only a small fraction, α, of the edges in any cut.

Figure 5 shows two examples. On the left (in blue) is a 1-thin tree, which is the fattest kind of tree possible. It contains *all* of the edges joining red vertices to black vertices, so the partition into red and black is a very bad cut. On the right (in

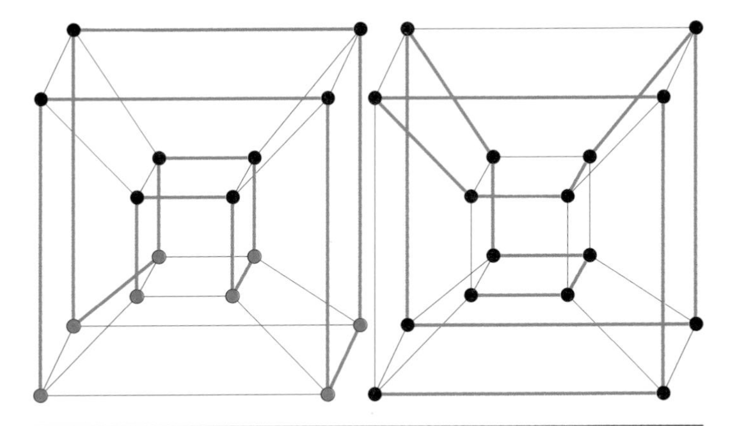

Figure 5. *A 1-thin tree (left) and a 2/3-thin tree (right). A 1-thin tree is actually the fattest possible. One algorithm to solve the ATSP requires alpha-thin trees, but their existence for large graphs is currently unproven. Oveis Gharan and Arani bypassed this requirement by using the Marcus-Spielman-Srivistava theorem. (Figures courtesy of Shayan Oveis Gharan.)*

blue) is a 2/3-thin tree. It can be shown, by a tedious analysis of all 32,768 possible partitions, that the blue tree contains at most 2/3 of the edges in any cut.

As this example suggests, α-spanning trees are hard to find and tedious to confirm. Luis Goddyn has conjectured that any graph that is sufficiently well connected has an α-thin spanning tree. But instead of waiting for Goddyn's conjecture to be proved, Oveis Gharan and Arani used a different idea. They imagined replacing the edges of the graph with resistors and sending an electric current through them. An *α-spectrally*-thin tree uses only a small fraction, α, of the energy. (Alpha-spectrally thin trees are always α-thin in the "usual," or combinatorial sense.)

The key point is that this energy function is *exactly the same kind of energy function* that appears in Weaver's Conjecture (see main article). It involves sums of *squares* of currents, rather than first powers. Thus they could adapt the Marcus-Spielman-Srivastava theorem to split the graph into two subgraphs, each with about half the energy. Then they split it again and again, eventually getting α-spectrally-thin trees with α as small as desired.

There was one further trick required. Remember that Santa can't apportion presents among two kids fairly (see main article) if one of them is too expensive. This corresponds to edges in the electrical circuit that suck up too much energy: a tree containing such an edge will not be spectrally thin. Anari and Oveis Gharan found an ingenious way around this by adding a very small number of shortcuts to dramatically cut the electric bill on the too-expensive bits of highway. Then they found an α-spectrally-thin tree on the modified highway system, using the Marcus-Spielman-Srivasta theorem. Finally, they showed that this tree could be used to produce a solution to the ATSP, not only on the modified graph but also on the *original* graph.

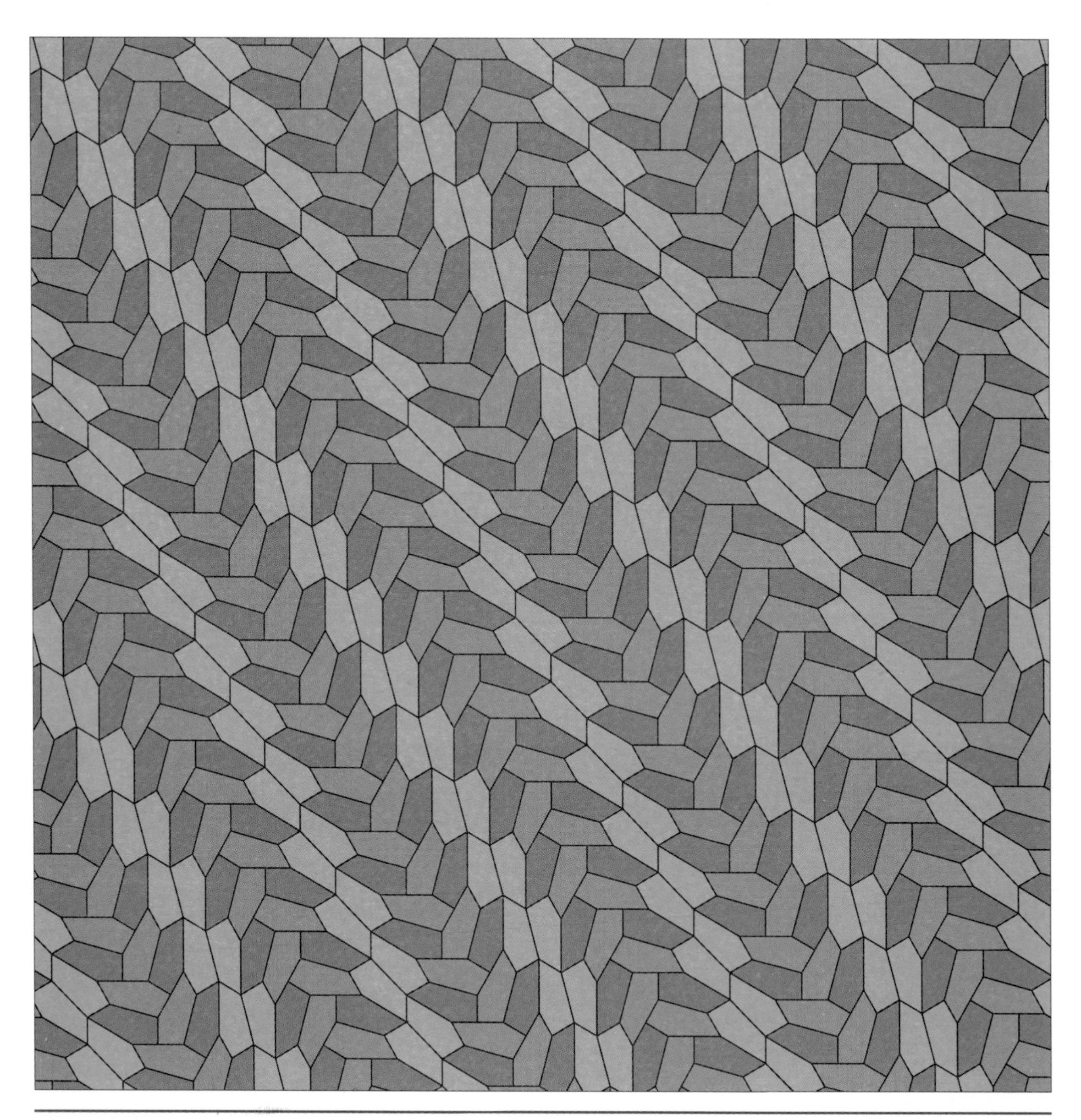

Five-sided Success. *A sample tiling by the latest addition to the list of convex pentagons capable of tiling the plane, discovered in 2015, by Casey Mann, Jennifer McLoud, and David Von Derau. The color coding indicates a property crucial to the pentagon's discovery: 3-block transitivity. (Figure courtesy of Casey Mann.)*

A Pentagonal Search Pays Off

Barry Cipra

IF YOU'RE LOOKING FOR A NEEDLE IN A HAYSTACK, it helps to be patient. If the needle and the hay are mathematical, it helps to have a computer as well. But even that may not be enough: Mathematical haystacks often contain an infinite amount of hay (and sometimes an uncountably infinite amount of the stuff!), so even when they also house an infinite number of needles, there's no guarantee you'll ever find one. You also need a smart way of organizing the search. And then it helps to be lucky as well. And observant.

All those elements came together last summer, when Jennifer McLoud, a mathematician at the University of Washington Bothell, reviewed the output of a computer run. She and her husband, Casey Mann, who is also a mathematician at UW Bothell, had spent a year and half developing an algorithm that would search a particular mathematical haystack. David Von Derau, an undergraduate math major, wrote a computer program implementing their algorithm to run on a high-performance computing cluster operated by the University of Washington. On July 27, while analyzing the output of Von Derau's program, McLoud noticed a needle. A new one that no one had seen before.

A fifteenth pentagon that tiles the plane.

The haystack that Mann and McLoud were examining is a mathematical problem with a rich history: the classification of

Casey Mann, Jennifer McLoud, and David Von Derau. *(Photo courtesy of Marc Studer - University of Washington Bothell.)*

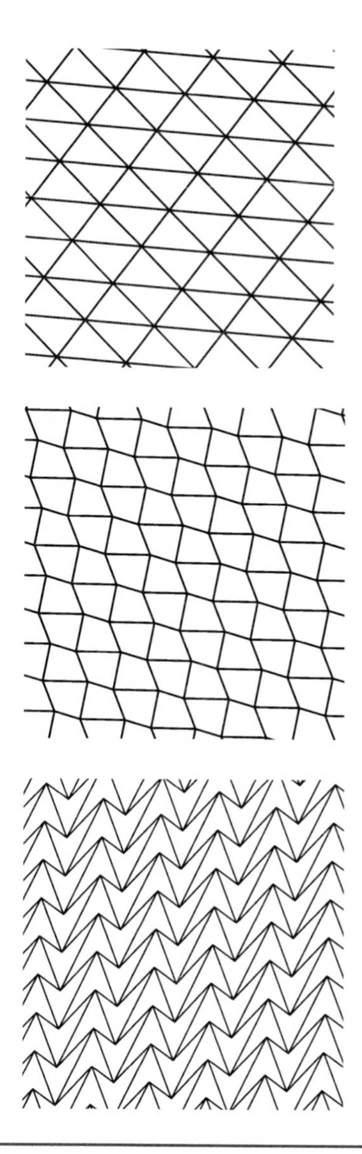

Figure 1. *Rotating a triangle around the midpoint of one of its sides produces a parallelogram, which easily tiles the plane (top). A similar rotation of a quadrilateral also tiles the plane (middle). The procedure even works for non-convex quadrilaterals (bottom). (Figure courtesy of Casey Mann.)*

convex polygons capable of tiling the plane—that is, covering the entire plane, without gaps or overlaps, using congruent copies of a single convex polygonal shape. The classification depends, of course, on the number of sides. For triangles and quadrilaterals the answer is simple: *Every* convex shape with three or four sides is capable of tiling the plane (see Figure 1). The answer is also simple for convex polygons with seven or more sides: *No* such shape is capable of tiling the plane. That leaves pentagons and hexagons.

On the one hand, five- and six-sided examples certainly exist. The regular hexagon is a familiar tiling, and it's easy enough to cut the regular hexagon in half to produce a convex pentagon that tiles the plane (see Figure 2). On the other hand, not every five- or six-sided polygon is capable of tiling the plane; for example, the regular pentagon cannot.

In 1918, the German mathematician Karl Reinhardt gave a complete classification of convex hexagons that tile the plane. He showed there are three families of such hexagons. He also found five families of convex pentagons that tile the plane. Each family contains infinitely many different tiles, and some tiles belong to more than one family—the regular hexagon, for example, belongs to all three hexagonal families. Membership in a family is determined by certain relationships being satisfied among the angles and edge lengths of a prospective tile (see Figure 3, next page).

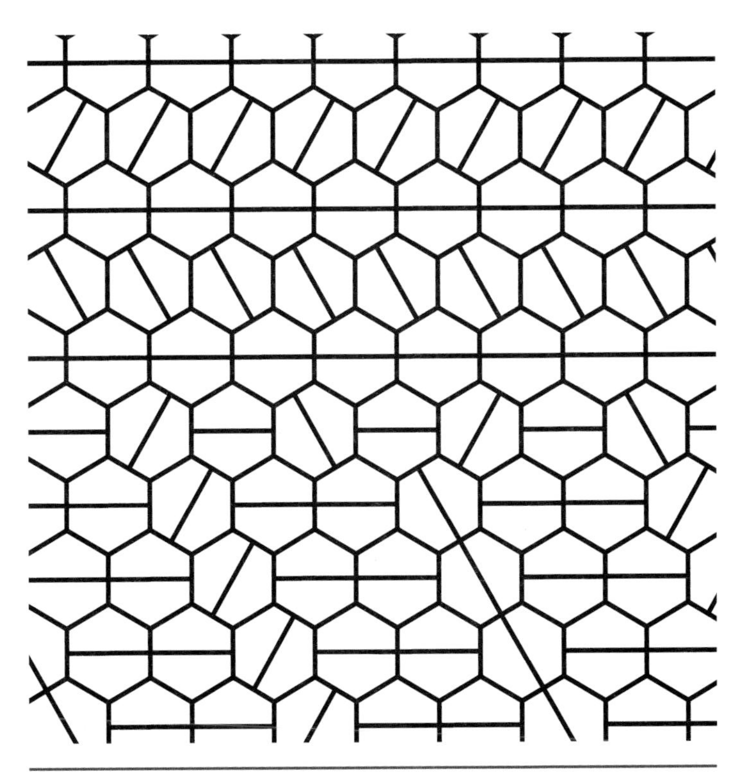

Figure 2. *If you start with a tiling by regular hexagons and cut each tile exactly in half, from the midpoint of one edge to the midpoint of the opposite edge, you get a tiling by pentagons. (It belongs to the family labeled Type 1 in Figure 6.) The choice of opposite midpoints can vary, so the resulting pentagonal tiling need not have any symmetries. Such tilings are called non-periodic. An open question: Is there a convex pentagon that* only *tiles the plane non-periodically? (Figure courtesy of Casey Mann.)*

The polygons in Reinhardt's families have a special property: Every one of them is capable not only of tiling the plane, but of doing so in such a way that there is a symmetry of the tiling that moves any given tile onto any other given tile. This "tile-transitive"—also known as "isohedral"—property looms large in the theory; for pentagons Reinhardt showed that any convex pentagon that can produce a tiling with the tile-transitive property belongs to one of the five families.

In the eighteenth of his famous twenty-three problems, laid out in a lecture for the International Congress of Mathematicians in 1900, David Hilbert had wondered about tile-transitivity in three dimensions: Was every polyhedron (convex or not) that is capable of tiling space necessarily also capable of doing so in transitive fashion? For some reason, he didn't bother asking the corresponding two-dimensional question. (Hilbert's 18th problem also included the Kepler Conjecture—see **"Quod Erat Demonstrandum,"** page 64 in this volume.) In 1928, Reinhardt answered Hilbert's question by producing an example of a space-tiling polyhedron that is not tile-transitive. Then, in 1935, Heinrich Heesch answered the unasked question: He found a ten-sided (non-convex) polygon that can tile the plane but only non-transitively.

This was pretty much where things stood until 1968, when Roger Kershner at Johns Hopkins University found three new families of convex pentagons that tile the plane. Kershner's examples did not have the tile-transitive property; indeed, they were the first examples of convex polygons that tile the plane but only non-transitively. (Curiously, in a 1974 symposium on Hilbert's problems, the Fields Medalist John Milnor, citing Heesch's ten-sided non-convex polygon in a talk on the eighteenth problem, seemed unaware of Kershner's pentagons: "I do not know whether a convex example exists," he wrote.

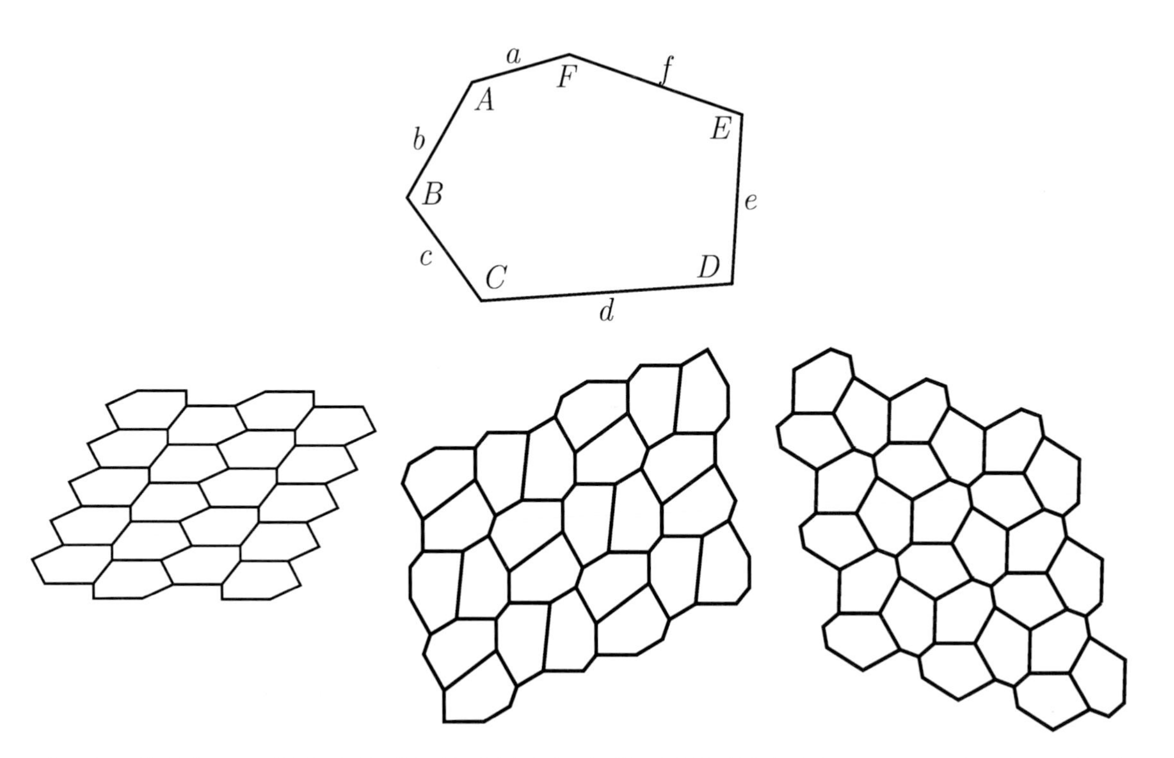

Figure 3. *There are three families of hexagons capable of tiling the plane, distinguished by relationships among the angles and side lengths of the tile, labeled as shown at top. The first family (left) satisfies the equations $A + B + C = 360°$ for the angles and $a = d$ for the sides. The second family (middle) satisfies $A + B + D = 360°$, $a = d$, and $c = e$. The third family (right) satisfies $A = C = E = 120°$, $a = b$, $c = d$, and $e = f$. The regular hexagon belongs to all three families. (Figure courtesy of Casey Mann.)*

Which may only show that the mathematical literature is its own enormous haystack.)

Kershner's paper, published in the *American Mathematical Monthly*, asserted that the three new families completed the list of convex pentagons capable of tiling the plane, but omitted the proof as "extremely laborious." The matter stood again until 1975, when the renowned mathematics writer Martin Gardner wrote about Kershner's result in his "Mathematical Games" column that July for *Scientific American*. Gardner's column prompted some remarkable correspondence.

First came a letter from Richard James III, a computer scientist at Control Data Corporation, with a pentagon that did not belong to any of the eight families Reinhardt and Kershner had identified. James's pentagon introduced a ninth family of tiling pentagons. Gardner reported the finding in his December column. And that's where things took off.

Marjorie Rice, a homemaker and mother of five whose son subscribed to *Scientific American*, saw Gardner's column and decided to see if there were any more pentagons that could tile the plane. An amateur with no mathematical background beyond high school, Rice invented her own approach to the problem, including pictorial notation to summarize the relationships amongst angles and side lengths for the nine known families (see Figure 4). Her "data storage" consisted of 3-by-5 index cards.

The approach paid off: in February, 1976, Rice wrote Gardner asking if a family of tiling pentagons she had discovered might be new.

Gardner passed the news along to Doris Schattschneider, a tiling expert at Moravian College in Bethlehem, Pennsylvania, who confirmed that Rice's discovery was indeed a new family

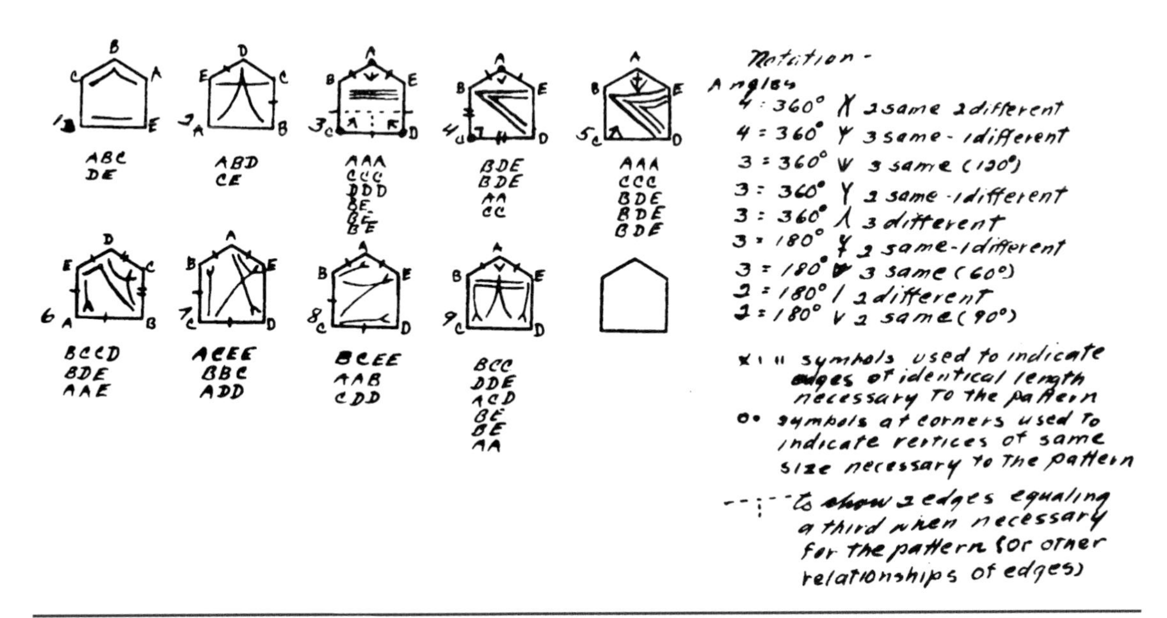

Figure 4. *Marjorie Rice invented her own notation to record pertinent information about pentagons that tile the plane: which angles meet at a vertex of a tiling (to add to 360 degrees) and which side lengths must be equal. Here is her inventory of the nine types that were known prior to her additional four discoveries. (Figure courtesy of the Rice family.)*

Figure 5. *Marjorie Rice's first discovery of a previously unknown pentagon that tiles the plane is the basis for a literal tiling of the seventh-floor elevator lobby at the Ohio State University Mathematics Tower in Columbus, Ohio. (Photo courtesy of Marilyn Radcliff.)*

(see Figure 5). "I asked Marjorie to write me all she had done on the problem and keep me informed of any new results," Schattschneider later wrote in an article, "In Praise of Amateurs," in a tribute book, *The Mathematical Gardner*. Rice did. Over the course of the next two years, she found another three families of convex pentagons that tile, bringing the list to thirteen.

The question arose: Was the list now complete?

It was not. In 1985, Rolf Stein, a mathematician at the University of Dortmund in Germany, found a new tiling pentagon that did not belong to any of the known families. This brought the list of known pentagonal families to fourteen (see Figure 6, page 94). Stein's pentagon was not only new, it was a true "needle": All of its angles and all of its side lengths (up to scale) are completely prescribed, so it belongs to a "family" of one.

And now, thirty years later, the list has grown by one again, thanks this time not to index cards and inspired guesswork but to a cluster of computers carrying out a methodical search.

The starting point for the Bothell researchers' work is the observation that the pentagons belonging to the nine known families beyond Reinhardt's, while never producing tile-transitive tilings in themselves, do allow for tile-transitive tilings by a larger (non-convex) polygon if you group two or three contiguous pentagons together. Tilings with this property are called 2- or 3-block transitive. More generally, if it takes a block of k tiles to produce a tiling that is tile-transitive, the original tiling is called k-block transitive, or k-isohedral.

Mann and McLoud's algorithm carries out an exhaustive search for k-block transitive pentagonal tiles for any given value of k. In a paper they are preparing for publication, they explain the mathematics underlying the algorithm. The basic idea is that there is a limited number of essentially different ways that a finite set of identical pentagons can be arranged next to one another. (The number would be severely limited if adjacent pentagons were required to share an entire edge; allowing for tilings that are not "edge to edge" is one of the complications of the theory.) The first trick is to systematically characterize the different arrangements, and to systematically explore what each arrangement implies for the relationships among the angles and side lengths of the constituent pentagon. For example, if the vertices of two copies of the pentagon meet along the edge of a third copy in a prospective 3-block, the angles at those two vertices must sum to 180 degrees.

The heart of the algorithm comes next: for each prospective k-block, see if it can tile the plane in transitive fashion. This becomes a finite calculation mainly because transitivity requires that each edge of the k-block polygon match up exactly (edge-to-edge) with another. The finite calculation, in principle, will methodically pick through a huge pile of pentagonal straw and report any needles.

"I began working on this problem, without much success, as early as 2007," Mann says. "In fall 2013, just after moving from Texas to Washington, Jennifer joined me in working on the problem, and we quickly developed the main ideas of the algorithm and developed a partial prototype that could handle the case when $k = 2$." Their prototype showed that the then-current list of tiling pentagons accounted for all 2-block transitive tilings.

"In winter 2014, we asked our then student David Von Derau, who was at the time a full-time software developer as well as a full-time student, if he wanted to help implement a more automated version of the algorithm," Mann says. "In spring 2015, after David had a beta version of his program, he ran it on the case $k = 3$ and sent us the output. At this point, we were in the process of debugging the code and optimizing it to run on the case $k = 4$. We didn't expect to find anything new in the $k = 3$ case."

Automated as it was, the beta version of the program didn't exactly have flashing lights to announce a needle. Instead, it winnowed the haystack down to a short list of possibilities that it was not equipped to analyze. These would need to be checked by hand, and would likely be exposed either as tilings already known or as "impossible" solutions which, while satisfying the algebraic equations of the problem, are not geometrically meaningful (such as "pentagons" that turn out to have a side whose length is negative).

On July 27, while sorting the wheat from the chaff, "Jennifer discovered one case that seemed to work," Mann says. He and McLoud quickly ascertained that it was indeed a geometrically meaningful pentagon, and that it was not any of the fourteen

The basic idea is that there is a limited number of essentially different ways that a finite set of identical pentagons can be arranged next to one another. (The number would be severely limited if adjacent pentagons were required to share an entire edge; allowing for tilings that are not "edge to edge" is one of the complications of the theory.)

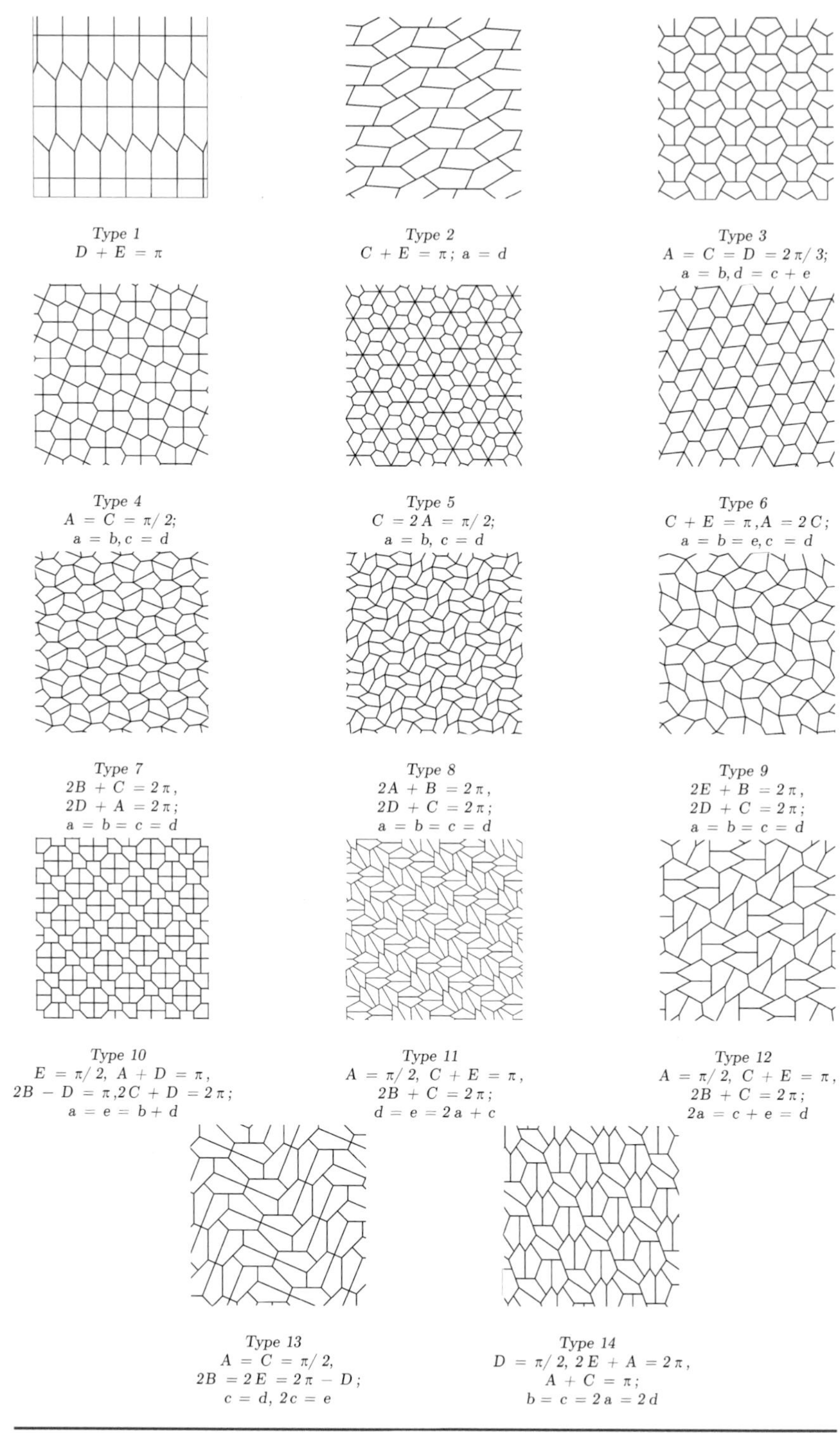

Figure 6. *The first fourteen families of tiling pentagons took seven decades to compile. The first types were identified in 1918; the next three in 1968. Type 10 was found in 1975, and types 9, 11, 12, and 13 in 1976–77. (The numbering is organized according to structural similarities in the relationships among the angles and side lengths for each type, as indicated by the equations in upper- and lower-case letters.) Type 14 was found in 1985. (Figure courtesy of Casey Mann.)*

known types. "Finally, as a visual check, I went to Adobe Illustrator and just made the tiling," Mann says. "The whole process probably took less than an hour."

Like Stein's 1985 pentagon, the new tile belongs to a "family" of one. Unlike Stein's its angles are all simple, familiar ones: 60, 135, 105, 90, and 150 degrees (see Figure 7). It's edge lengths are similarly simple. The new pentagon can, in fact, be cut into four pieces with familiar shapes: two right isosceles triangles (of different sizes) and two 30-60-90 triangles (also of different sizes).

Mann and McLoud plan to implement their algorithm to cover the cases $k = 4$ and $k = 5$. Beyond that, they say, the algorithm appears impractical for current computers. Even if it becomes practical, their algorithm alone will never definitively complete the classification of convex pentagons that tile the plane, because it proceeds one k at a time. In that regard it's comparable to another ongoing computer effort, the Great Internet Mersenne Prime Search (GIMPS), which has computers around the world examining numbers of the form $2^n - 1$ to see if any of them are prime. Every so often, GIMPS reports the sighting of a new Mersenne prime; much more often it rules out entire stretches of exponents n that produce only composite numbers.

Unlike GIMPS, though, Mann and McLoud's algorithm could conceivably overlook a needle even if it ran forever. That's because the tilings it looks for are necessarily periodic: they all have translational symmetry, like wallpaper patterns. It's conceivable there might be a pentagon somewhere in the haystack that is capable of tiling the plane but only in non-periodic fashion. It seems reasonable to believe that no such pentagon exists, but at one time it seemed reasonable to believe that Reinhardt's, then later Kershner's, lists were complete. In mathematics, as long as there's a haystack, there's a possibility of needles.

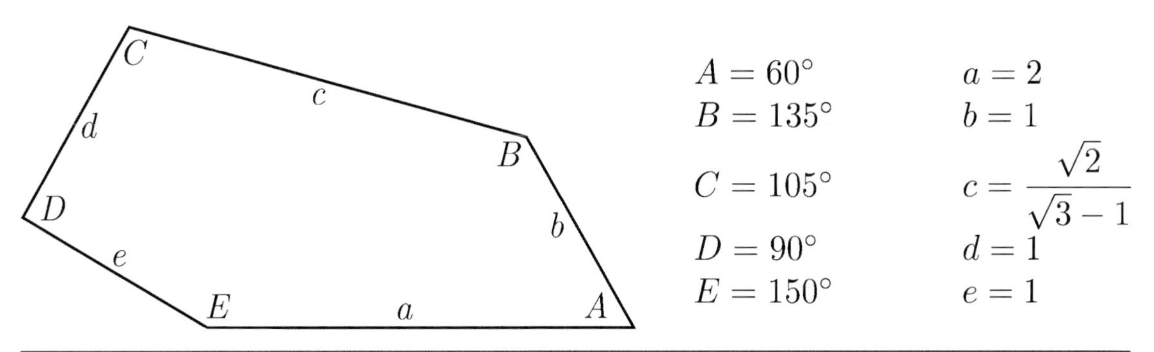

Figure 7. *The newest addition to the list of tiling pentagons, discovered by Casey Mann, Jennifer McLoud, and David Von Derau, even looks a bit like a (fat) needle. If you add line segments from E to B and E to C and a perpendicular from E to BC, it decomposes into four familiar triangles. (Figure courtesy of Casey Mann.)*

Mike Trout (left) and Miguel Cabrera (right) *caused furious debates among baseball fans in 2012 over who deserved to win the Most Valuable Player award. "Sabermetrics" or "analytics" have dramatically changed the way that players are evaluated and even the way the game is played on the field. (Photo of Mike Trout courtesy of Debby Wong/Shutterstock.com and the photo of Miguel Cabrera courtesy of Photo Works/Shutterstock.com.)*

The Brave New World of Sports Analytics

Dana Mackenzie

AUGUST 4, 2015. TOP OF THE NINTH inning, no outs, man on first base. Luis Valbuena of the Houston Astros hits a weak ground ball to the Texas Rangers' shortstop. Except... the shortstop isn't there. In a designed play, called a "shift," the Rangers' manager, Jeff Banister, had moved him to the right-hand side of the diamond because the statistics said that Valbuena was more likely to hit the ball that way. (See Figure 1.)

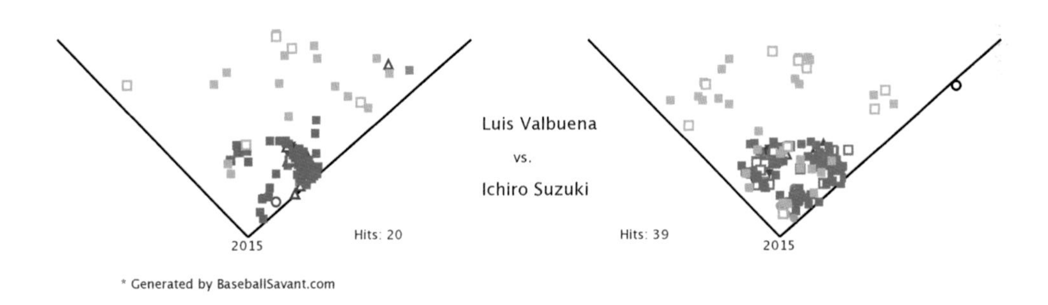

Figure 1. *A spray chart shows where every ball hit by a batter in 2015 has landed, and the outcome (out = red; base hit = green; double play = blue). Luis Valbuena tends to hit to the right-hand side of the field, and opponents can take advantage by positioning fielders on that side. Ichiro Suzuki, one of the most skilled hitters in baseball, hits in all directions. (Figure courtesy of Daren Willman.)*

Harold Reynolds, a former player commenting for the Major League Baseball Network, hates the shift. "That's been a double play ball ever since Abner Doubleday!" he exclaims. If the Rangers' shortstop had been playing in his normal position, he would have caught the ball and his team would likely have gotten two outs (a double play). If so, the Astros would have had only a 6.7 percent chance of scoring a run and tying the game. Now, with runners on first and third base, the Astros have an 87 percent chance to tie or go ahead.[1]

[1]Nevertheless, the Rangers' pitcher saved the day by retiring three Astros batters in a row without allowing the runner on third base to score. In all likelihood, his performance saved Banister from having to answer some pointed questions about his defensive strategy.

Keith Woolner. *(Photo courtesy of Dan Mendlik/Cleveland Indians.)*

The double play that didn't happen is just one tiny example of a sea change that is occurring not only in baseball, but in nearly all professional sports: the rise of what is known variously as statistics, big data, or "analytics." In baseball, newly designed measures of value affect trades, contract negotiations, and Most Valuable Player awards. In basketball, analytics redefine shot selection and provide new metrics for defensive performance. In both of these sports, an even greater analytical upheaval is on the horizon, thanks to the recent introduction of video cameras that monitor every player, every second (in fact, 25 times a second).

Even if you aren't a sports fan, it's worth paying attention because sports illustrate what can (and cannot) be done with "big data" under optimal conditions. "There's a perfect storm of circumstances in sports that makes rapid analytical progress possible decades before other fields," wrote Nate Silver in *ESPN: The Magazine*. Silver, a prominent statistician who specializes in sports and politics, argues that sports have not just big but high-quality data: impartial, long-term and meticulously collected. Also, sports have well-defined rules. And third, they have clear benchmarks of success. Constant competition winnows out bad ideas in a hurry. If analytics can't succeed in sports, then they will never succeed in business, government, economics, or politics.

Even so, they faced an uphill battle. Traditions die hard, as Reynolds' comment about Abner Doubleday (the legendary inventor of baseball) shows. Given a rational statistical solution to a problem, people will often default to simpler rules of thumb, even if they are demonstrably inferior. The public has trouble accepting that statistically sound decisions may backfire because of chance factors. And the media love to bash the "geeks" who do sports analytics for a living. "They're a bunch of guys who have never played the game, and they just want to get in the game," said Charles Barkley, a Hall of Fame basketball player turned television personality, in a 2015 interview.

While fans and ex-stars ridicule the statisticians in public, in the privacy of team offices it's a different story. One team after another is showing its support for analytics in the most concrete way: by hiring analysts. "In the beginning, if there was a role for somebody like me, it would be one person per team," says Keith Woolner, the director of baseball analytics for the Cleveland Indians. "Now teams are hiring more and more analysts, three or four or five per team, and I hear rumors that one team has eight. It's become a much more entrenched part of front office operations. That, in the long term, has been the biggest surprise to me."

The Era of Sabermetrics

Woolner is probably the kind of guy Barkley was talking about in his infamous rant about analysts. Woolner grew up in New England, rooted for the Boston Red Sox, played Little League baseball, and collected baseball cards as a kid. When he went to college at MIT he majored in computer science and applied mathematics, but he still retained his interest in baseball statistics.

"Even before the World Wide Web existed, I participated in a discussion forum called rec.sports.baseball," Woolner says.

"That was my first exposure to the emerging sabermetric point of view."

"Sabermetrics" came into being in the late 1970s and early 1980s, when an iconoclast named Bill James began publishing the *Bill James Baseball Abstract*. It started small but became a huge hit among baseball fans. James subjected all of baseball's time-honored traditions to scientific analysis. With no outs and a runner on first base, should the manager call for a bunt? (This is a traditional strategy where the offensive team gives up an out in order to advance the runner to second base.) Does the batting average, a traditional measure of prowess for hitters, really show who helps their teams the most? Prove it! Time after time, James (and other pioneers) showed that conventional wisdom could not stand up under even the simplest of statistical analyses. The Society for American Baseball Research (SABR) became a haven for like-minded people, and lent its name to the scientific approach known as sabermetrics.

Suddenly, people with math or statistics or computer science degrees were prized commodities, and they didn't have to be baseball insiders.

Throughout the 1980s and 1990s, sabermetrics was primarily the domain of baseball outsiders. That began to change around the turn of the century. Under general manager Billy Beane, the Oakland Athletics assembled one of the best baseball teams in the major leagues by defying conventional wisdom. For example, they looked for players with a good on-base percentage (OBP) instead of a good batting average. It was a move straight out of the Bill James playbook. For four straight years, they made the playoffs despite having one of the lowest payrolls in baseball, by recognizing the value in players whom other teams didn't want. That, at least, was the spin that writer Michael Lewis put on their success in an influential book called *Moneyball*, which eventually became a Hollywood movie.

When the Boston Red Sox won the World Series in 2004, even more teams took notice. The Red Sox' general manager, Theo Epstein, was a convert to sabermetrics who had hired Bill James as a consultant in 2002. At that point, the Red Sox had not won a World Series in 84 years. Two years later, they were champions. Coincidence? Probably. But in baseball, where everybody imitates a champion, it opened the floodgates.

Suddenly, people with math or statistics or computer science degrees were prized commodities, and they didn't have to be baseball insiders. In 2002, the Toronto Blue Jays hired Keith Law, a writer for *Baseball Prospectus* (a successor to the *Bill James Baseball Abstract*). In 2004, the Athletics plucked Farhan Zaidi, an economics graduate student, from a pile of more than a thousand resumes. By 2007, when Woolner decided to leave his day job as a software developer, it took him only two months to land a job as an analyst for the Cleveland Indians.

Woolner already had a long track record as a writer for *Baseball Prospectus*. In 2000 he published an article on the "Hilbert problems" of baseball, imitating the mathematician David Hilbert, who in 1900 made a list of the 23 most important math problems for the next century. (Sample problem: How do you measure the catcher's value for run prevention?) However, his best-known innovation was Value over Replacement Player (VORP), a measure of a player's total contributions, both positive and negative, in all aspects of the game (hitting, fielding, running, and pitching if applicable). Woolner realized that if a player were injured, his team would have to replace

Ben Baumer. *(Photo courtesy of Eric Kilby.)*

him not with an average major-league player, but an average *replacement* player, typically a minor leaguer performing below major-league standards. Thus the comparison to a replacement player, not to an average player, defined his true worth.

Another version of VORP, called Wins Above Replacement (WAR), has become perhaps the most widely discussed new-age baseball statistic. It raised a controversy in 2012, when the vote for Most Valuable Player (MVP) in the American League became a two-man contest between a rookie, Mike Trout of the Los Angeles Angels, and a veteran, Miguel Cabrera of the Detroit Tigers. Trout was a great hitter, a brilliant fielder, and an outstanding baserunner. He contributed 10.8 wins to his team, the best Wins Above Replacement score ever amassed in the short time since that statistic had been invented. Cabrera was a poor fielder and a pedestrian baserunner, but had a historic season as a hitter. He was the first player in 45 years to win the Triple Crown, leading the league in the three traditional categories of home runs, runs batted in, and batting average. But according to WAR his value to his team was only 7.2 wins, and he was not even the second-best player in the league.

The 2012 MVP race became a holy war between traditionalists and sabermetricians, which the traditionalists (and Cabrera) won. However, Trout may get the last laugh: He won the MVP award in 2014, and is widely considered the best player in baseball today.

One little-noticed irony is that WAR, in spite of being a "scientific measure" of playing ability, actually falls far short of scientific standards. As Ben Baumer, a statistician at Smith College, points out, WAR lacks a measure of uncertainty and it is irreproducible, in the sense that the casual fan cannot compute it from publicly available data.

The lack of transparency is a problem throughout sports analytics. Teams jealously guard their secrets. Even Woolner, the inventor of VORP, is not at liberty to say whether the Indians use WAR or VORP or something like it. Three separate companies (Fan Graphs, Baseball Prospectus, and Baseball Reference) sell their own versions of WAR, all using proprietary formulas. "They'll write a blog post about it, but some of these posts you can follow and reproduce the right numbers, while others you can't," Baumer says.

In 2015, Baumer, together with Shane Jensen of the Wharton School of Business and Gregory Matthews of Loyola University of Chicago, introduced a new "openWAR" that makes the calculation transparent and brings it up to 21st century scientific standards. In fact, it's a beautiful example of how statistical modeling can go beyond the mere tabulation of data. For every play (such as Valbuena's ground ball to shortstop) they compute the change in the offensive team's run expectancy as a result of that play. The responsibility for that change is apportioned, using linear regressions on the entire database of plays for the season, into ballpark effects (certain parks are easier to hit in than others), platoon effects (generally, hitters do somewhat better against opposite-handed pitchers), the hitter's contribution, and the baserunners' contribution. Because they are statistically derived, each of these contributions has some uncertainty.

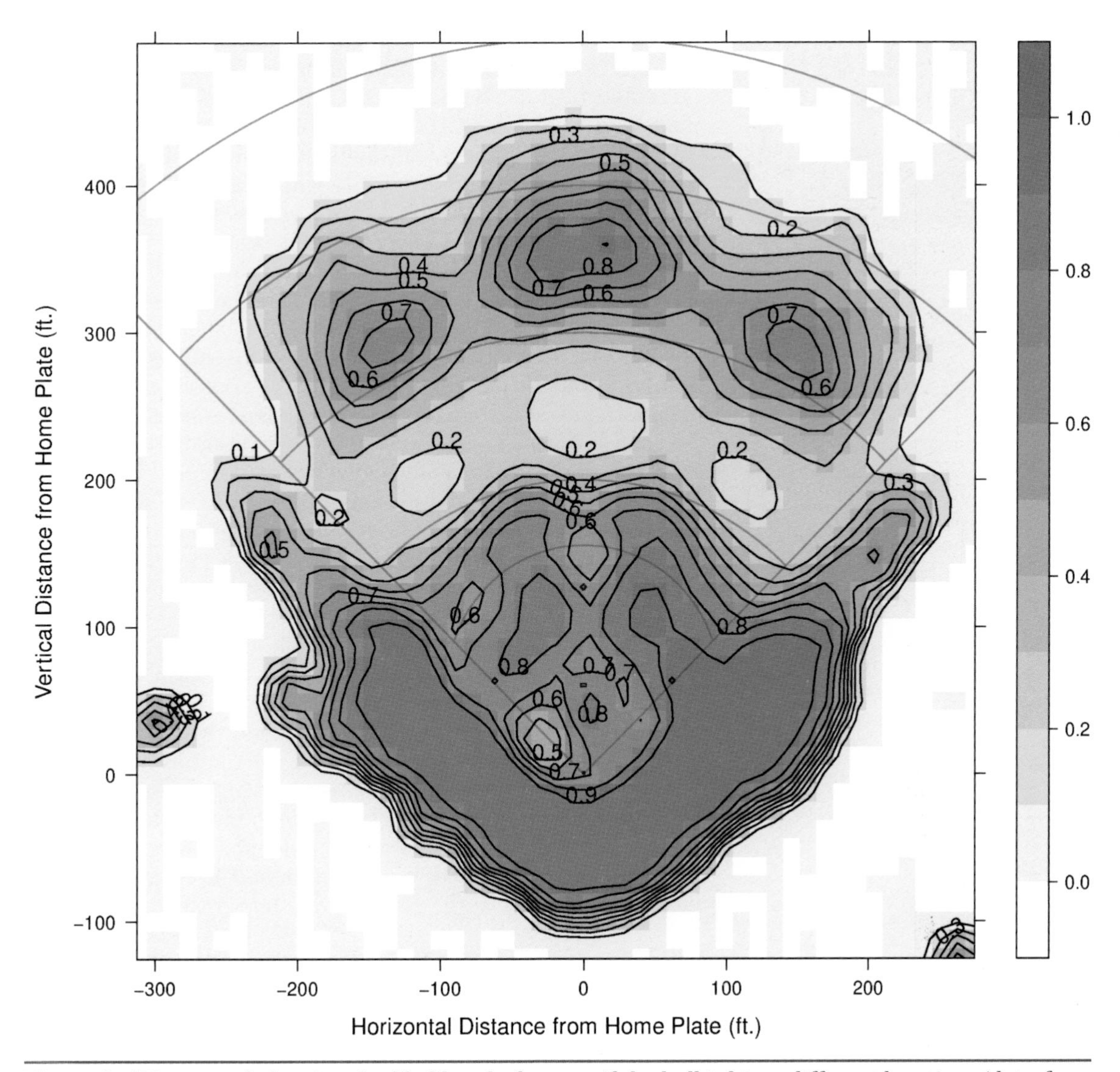

Figure 2. *"Heat map" showing the likelihood of an out if the ball is hit to different locations (data from 2013). (Figure courtesy of Ben Baumer.)*

On the defensive side, the apportioning of credit or blame is quite a bit more interesting. Baumer, Jensen, and Matthews made a contour map of the likelihood of an out, given the spot where the ball landed (see Figure 2). Any deviation from that expectation is either the pitcher's fault (he let the batter hit the ball "where the fielders ain't") or the fielder's fault (the batter hit the ball to a place where the fielder should have gotten to it). Neither the pitcher's nor the fielder's contributions can be estimated by linear regression. For fielding, it's a logistic regression (i.e., a regression using quadratic terms), and for pitching it's a "non-parametric kernel-smoothed model," Baumer says. But if you don't like the way he did it, you can plug in a different statistical model. That's the beauty of open source code.

Finally, each player's run contributions to every play throughout the season is summed up and converted into an equivalent number of wins. Baumer's group also estimated the uncertainty in the openWAR coefficient. The lack of uncertainty estimates is an egregious omission in most sports statistics. In a traditional measurement like batting average, there is no uncertainty. If a player gets 30 hits in 100 at-bats, his batting average is .300, period. Yet batting averages fluctuate greatly from year to year, for reasons that may have to do with chance or may be due to genuine changes in the player's ability. If your star hitter batted .300 last year but only .275 this year, is he "in a slump" or "over the hill," or is it just chance variation? Traditional baseball statistics give no guidance, because they do not measure uncertainty.

By contrast, the openWAR code allowed the researchers to plot a *probability distribution* for a player's possible WAR coefficients. Baumer's group replayed simulated versions of the 2012 season 3500 times, and showed that the overlap between Miguel Cabrera and Mike Trout was larger than one might expect. In 31 percent of the simulated seasons, Cabrera had the higher WAR. So, looking at it optimistically, there is only a 69 percent chance that the wrong player won the Most Valuable Player award. But in 2013, Trout had a more pronounced advantage (see Figure 3) and still lost.

Baumer, by the way, is an example of a baseball analyst who went in the opposite direction from the norm. From 2004 to 2012 he worked in baseball for the New York Mets, while studying for a Ph.D. at the same time, and then he took a job in a university math department. Clearly Charles Barkley was wrong that all sports analysts are just tying to horn in on the athletes' fun. Some of them actually prefer the greater openness and lesser day-to-day pressure of academia. "When you work for a team, things need to happen immediately, and it's very easy to lose track of the goals of long-term research," Baumer

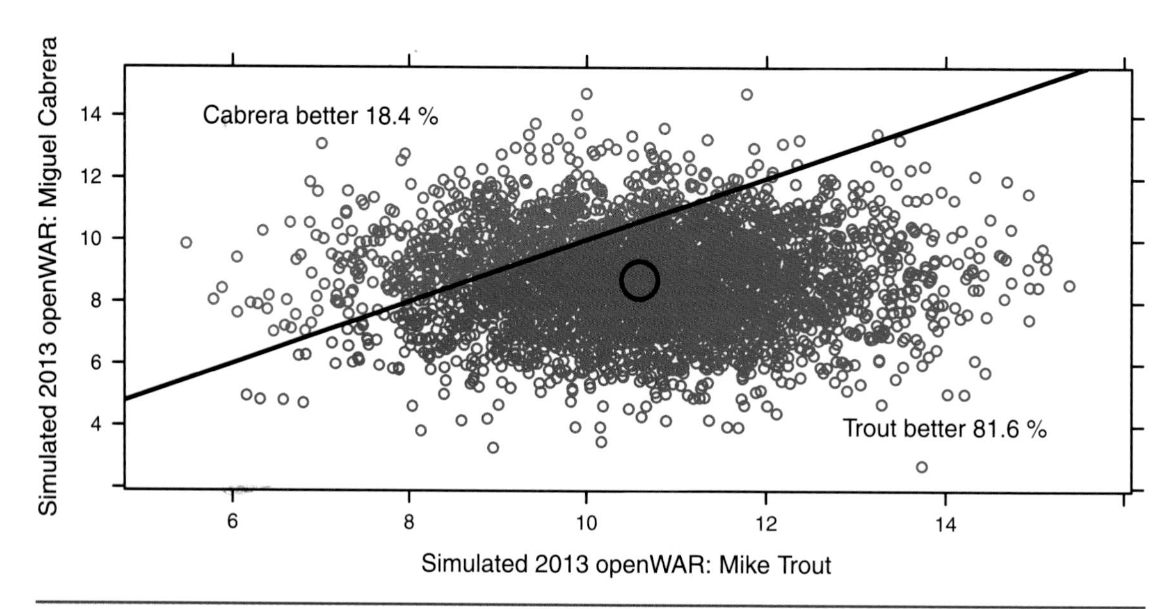

Figure 3. *Miguel Cabrera's value versus Mike Trout's value, measured in terms of Wins Above Replacement in 3500 simulated versions of the 2013 season. The disparity between the two was even greater than in 2012, but Cabrera won the Most Valuable Player award again. (Figure courtesy of Gregory Matthews and Ben Baumer.)*

Figure 4. *Pitch framing. A catcher convinces an umpire that a borderline pitch is a strike. Proficient catchers can "steal" about two strikes a game, and this may add up to 3-4 extra wins for their team over the course of a season. The importance of pitch framing could only be guessed at before the age of analytics and video-captured data. (Photo courtesy of Brian McEntire/istock.com.)*

says. "It's much more possible in academia to let an idea take two or three years to come to fruition."

In what other ways have analytics affected the game of baseball? A very interesting, and recent, case study is the evaluation of catchers—problem #3 on Woolner's "Hilbert problems" of baseball. There is a great deal of lore in baseball about the importance of the catcher, the team's "field general" who calls all the pitches. At first sabermetricians thought this value was overrated, and it was yet another point of contention between them and traditional baseball folk.

However, in 2006 a major new source of data became available: PITCHf/x, a camera-based system marketed by Sportvision that records the location and speed of every pitch. For the first time sabermetricians could objectively compare the location of a pitch with the umpire's call. In baseball, an umpire behind the plate rules every pitch a ball or a strike, a judgement that is in principle clearly defined by the rulebook. A pitch is in the "strike zone" if it passes over home plate at a height between the batter's knees and armpits. In practice the strike zone depends very much on the umpire's eye, and it can even vary from day to day.

As it turned out, the pitch call is also strongly affected by the catcher. "Pitch framing" is a skill that has been talked about for years, but previously there had been no way to measure its importance. It is the art of catching the ball (the catcher's #1 job!) while keeping the catcher's glove in a place that makes it look as if the pitch was in the strike zone (see Figure 4). Analysts Mike Fast, Max Marchi, Harry Pavlidis and Dan Brooks showed

Baseball Glossary

For readers who have escaped indoctrination into the Church of Baseball (as it was referred to in the 1988 movie, *Bull Durham*), here is a brief explanation of relevant terms.

Strike. A pitch within the "strike zone," which counts against the hitter if he lets it go by. If a batter accumulates three strikes, he strikes out.

Ball. A pitch outside the strike zone, which counts in the hitter's favor if he lets it go by. If a batter accumulates four balls, he is awarded first base (a base on balls, or "walk").

Out. A strikeout, or a batted ball that the defensive team catches on the fly, or a batted ball that the defensive team catches on the bounce and throws to first base before the batter gets there. After the hitting team accumulates three outs, it loses the chance to score any more runs and the other team gets a chance to hit.

Base hit. If the batter hits the ball and reaches first base (or farther) without being put out, he is credited with a base hit.

Run. When a player touches all four bases, including home plate, without being tagged out or thrown out, he scores a run for his team. The team that scores the most runs wins.

Traditional Measures of Batting Skill (not based on sabermetrics)

Batting average. Base hits divided by at-bats. Flaw: Since 1887, a walk has not counted either as a hit or as an at-bat. Yet the ability to draw walks is an important part of hitting ability, especially because opponents sometimes walk dangerous hitters intentionally, giving them one free base, to avoid the risk that they might hit a home run (four bases).

in *Baseball Prospectus* that a good catcher could "steal" a couple of strikes a game on borderline pitches. Pavlidis and Brooks estimate the value of a stolen strike at about one-ninth of a run. It followed that over a 162-game season, a catcher could save his team 30 or 40 runs, worth roughly three or four wins in WAR terms.

Several teams noticed. Fast was hired by the Houston Astros. Marchi was hired by the Cleveland Indians. "It's so obvious that teams are pursuing this when you see catchers that clearly don't hit or run well getting multi-year contracts," says Baumer. "To the teams that aren't hip to this, it's probably baffling. Jose Molina was signed to a 2-year deal with the Rays, even though he was basically the worst offensive player in baseball. The Pirates signed Russell Martin to a multi-year deal, because they figured out that he has this ability."

What would happen if analysts had access to video records not only of each pitch as it crossed home plate, but everything that happened on the field? This brave new world arrived in 2015, when Major League Baseball installed a video system called StatCast in every ballpark. Analysts can now measure the speed of the ball when it leaves the bat and how long it spends in the air. The new video data could revolutionize the evaluation of hard-to-quantify aspects of the game like defense and baserunning... if the teams can figure out how. "Based on

Home run. A base hit in which the batter touches all four bases without being thrown or tagged out. Home runs are a traditional measure of a hitter's "power." Flaw: Because the number of home runs is an absolute number, not an average, a hitter who has more at-bats has an advantage over one who has fewer.

Runs batted in. When a hitter gets a base hit (or a walk, or even sometimes an out) that directly causes a runner to score, the hitter gets credit for a "run batted in" or RBI. Flaws: Inconsistent with the definition of batting average. Highly dependent on the situations when a hitter happens to come to bat. Based on the theory that some players are better at hitting "in the clutch," which has very little scientific support.

Sabermetric Measures of Batting and Other Skills

On-base percentage (OBP). Number of times reaching base divided by plate appearances. The main difference between OBP and batting average is that walks are counted, in both the numerator and denominator. Though invented before the sabermetric era, OBP only became really popular in the 1980s and later.

Value over Replacement Player (VORP). A metric proposed by Keith Woolner that attempts to measure the total number of runs that a player contributes to his team in all aspects of the game: running, hitting, fielding, and (for pitchers) pitching.

Wins Above Replacement (WAR). Later versions of VORP that estimate the fractional number of *wins* that a player contributes to his team. Measures of running and fielding have become more sophisticated, but they are also proprietary: an ordinary fan cannot use publicly available data to calculate a player's WAR.

conversations I had six months ago, it sounds as if teams are at the very early stages of working with that data," Baumer says. "It's orders of magnitude more than they are used to."

To Woolner, StatCast is like the opening of a new treasure chest. "Every time we've had access to a new data set that told us a level of detail or depth beyond what we had before, there have been astonishing advances and insights that came out from that."

In Defense of Defense

Other professional sports do not have quite as long a tradition of statistics as baseball does, but they are catching up fast. If 2003 was the year of *Moneyball*, 2014 was the year of "money puck," as several professional hockey teams hired analysts. "Hockey has done a 180- degree turn in just the last two years," says Michael Lopez, a mathematician at Skidmore College. "Now about twenty out of thirty teams have analysts."

The premier American spectator sport—American football—has been the slowest to adopt analytics. It's not for lack of opportunity, or importance. For example, the National Football League instituted a significant change in the rulebook for the 2015 season: the extra point, a 1-point kick awarded after an offensive team scores a touchdown, will now be taken from the 15-yard line instead of the 2-yard line. The goal is to make one of the most boring, routine plays in football a little bit more exciting, with a greater risk of failure. It may also encourage teams to "go for two" more often. Teams can still elect to run or pass the ball from the 2-yard line, earning two points instead of one if they get into the end zone. In the past, two-point attempts have been rather rare except in specific situations, because teams preferred the "sure thing" of kicking the ball for one point.

The rule change could have been an ideal opportunity for statistical analysis: what would teams' optimal strategies be under the new rule? Where should the kick be taken from to create the optimal balance of risk and reward? Should the line of scrimmage for a passing or running attempt be changed, too? But there is no evidence that the league conducted any such analysis. "I think that they just picked the 15-yard line because it was a round number," Lopez says.

Somewhere between baseball and football on the learning curve is basketball. While baseball got StatCast in 2015, professional basketball's version, called SportVU, was installed in every team's stadium in 2013. So far, teams are quite closed-lipped about what they are doing with the data. However, a team of researchers at Harvard University called XY Hoops (now renamed XY Sports) has provided an exemplary model.

XY Hoops group at Harvard. *From left to right: Andrew Miller, Alex Franks, Alex D'Amour, Luke Bornn and Dan Cervone. (Photo by Kris Snibbe and courtesy of Harvard Public Affairs & Communications.)*

XY Sports was the brainchild of Luke Bornn, who specializes in spatial statistics, and Kirk Goldsberry, a specialist in geography. They were able to negotiate access to the SportVU data from Stats LLC, the company that owns it. In 2015 they published a paper, "Characterizing the Spatial Structure of Defensive Skill in Professional Basketball," that is almost a primer on spatial statistics. It also shines a light on one of the most difficult things to quantify in basketball (as it is in most sports): defensive performance.

To begin with, Bornn and Goldsberry ask the question: how can we tell from the video data who is guarding whom? A naive answer might be that each defensive player is guarding the offensive player closest to him. But that doesn't work, according to one of the XY Hoops graduates, Alexander Franks (now at the University of Washington). "If you watch an animation, you'll see the defensive assignments flipping like crazy all the time." (See Figure 5.)

Another solution to the problem would be to have experts watch the videos and identify the defensive matchups. But that would be labor intensive and it would preclude the possibility of actually learning something new that the experts don't know.

The solution adopted by XY Hoops involved two main ingredients: the expectation maximization (EM) algorithm and a hidden Markov model (HMM). They expressed each defensive player's location μ in barycentric coordinates, as a combination of the positions of the offensive player on the other team (O), the basketball (B), and the hoop (H):

$$\mu = \gamma_1 O + \gamma_2 B + \gamma_3 H \quad (\gamma_1 + \gamma_2 + \gamma_3 = 1).$$

The parameters $\gamma_1, \gamma_2, \gamma_3$ are unknown and have to be inferred

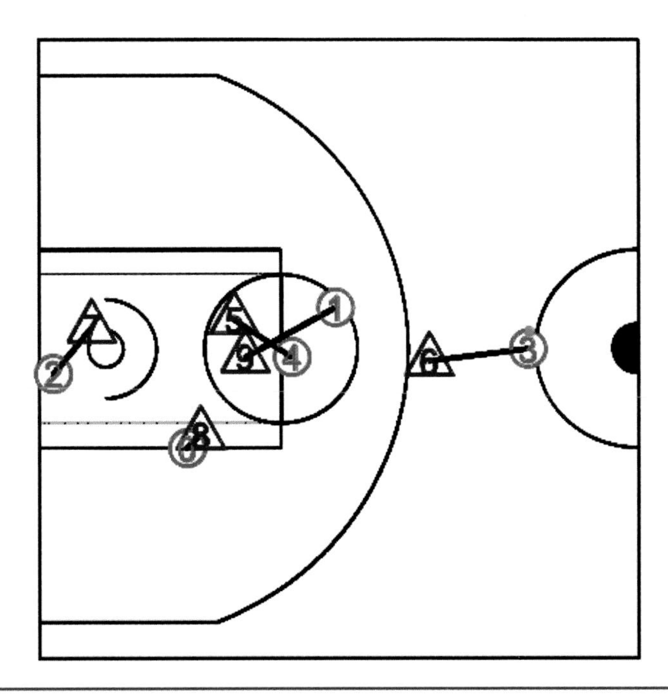

Figure 5. *Miller and Franks' algorithm infers, without human input, who is guarding whom in a basketball game. Here the algorithm deduced that 9 is guarding 1, even though 9 is closer to 4 at the moment. Video confirmed that 9 was, in fact, chasing 1 across the court. (Figure courtesy of Institute of Mathematical Statistics, from "Characterizing the spatial structure of defensive skill in professional basketball," by Alexander Franks, Andrew Miller, Luke Bornn, and Kirk Goldsberry,* Annals of Applied Statistics, *2015, Vol. 9, No. 1, pp. 94–121.)*

The HMM assumes that defensive switches are relatively uncommon, and that they occur at random times. Any given player might have, say, a 4 percent chance of switching his assignment every second. "This assumption makes the assignments more fluid over time, so you don't get the crazy switching," Franks says. "It was cool to see how nicely the animations worked."

from the data. Goldsberry and Bornn assumed that the actual position of the defender is determined by a "canonical defensive location" that they are trying to occupy, added to a certain amount of random variation whose magnitude is unknown. They also assume that if a defensive player is not guarding player O, then y_1 is meaningless; a player only aims for an ideal position with respect to the player he is guarding. However, the identity of that player is unknown, and given by an "indicator variable" that also has to be inferred from the data. (The indicator is 1 for the player the defender is guarding, and 0 for the players he is not guarding.)

"The EM algorithm is a tool for filling in missing data," Franks explains. "Here what is missing is the defensive assignment. The EM algorithm assigns probabilities to that missing value. There are two steps. First is the expectation—you take an expected value of who you're most likely guarding, given the current estimates of the canonical location [i.e., the gammas]. The next step, given those estimates, is to update the parameters [the gammas] using maximum likelihood." The EM algorithm is repeated ten or twenty times until it converges to an answer. But the results still don't pass the "eye test". They're too jumpy.

That's where the second ingredient, the hidden Markov model, comes in. The HMM assumes that defensive switches are relatively uncommon, and that they occur at random times. Any given player might have, say, a 4 percent chance of switching his assignment every second. "This assumption makes the assignments more fluid over time, so you don't get the crazy switching," Franks says. "It was cool to see how nicely the animations worked."

Already, this step of their model gave the XY Hoops team information that nobody ever knew before. They learned that the canonical defensive position is about two-thirds of the way between the offensive player and the basket, shading a little bit (roughly ten percent) toward the current position of the ball if it is held by a different player.

But figuring out who a player is guarding is only part of the process of determining defensive efficiency. The next challenge is to figure out what kind of shot the offensive player likes to take. Here, Franks' co-author Andrew Miller proposed using a nonnegative matrix factorization (NMF) model. This model supposes that there are a small number of shot types: for instance dunks, layups, mid-range jumpers, 3-pointers from the corner, 3-pointers from the top of the key (see Figure 6). But again, instead of having someone watch the video and identify the shot type, Miller's goal was to let the data itself speak.

In a nonnegative matrix factorization, each player has a column vector that summarizes his shot frequency from every position on the court. This data matrix, Λ, is factored into two matrices $\mathbf{L}$ and $\mathbf{W}$ that correspond respectively to shot types and player preferences. When Miller used the five shot types described above, the matrix $\mathbf{L}$ would have five columns, each one of which can be visualized as a "heat map". For example, dunks are always taken from very close to the basket and never taken from long-range, so the heat map for a dunk would be bright red near the basket and blue farther away. In numerical terms, each of the five columns would contain 2350 entries, one for each square foot on the court.

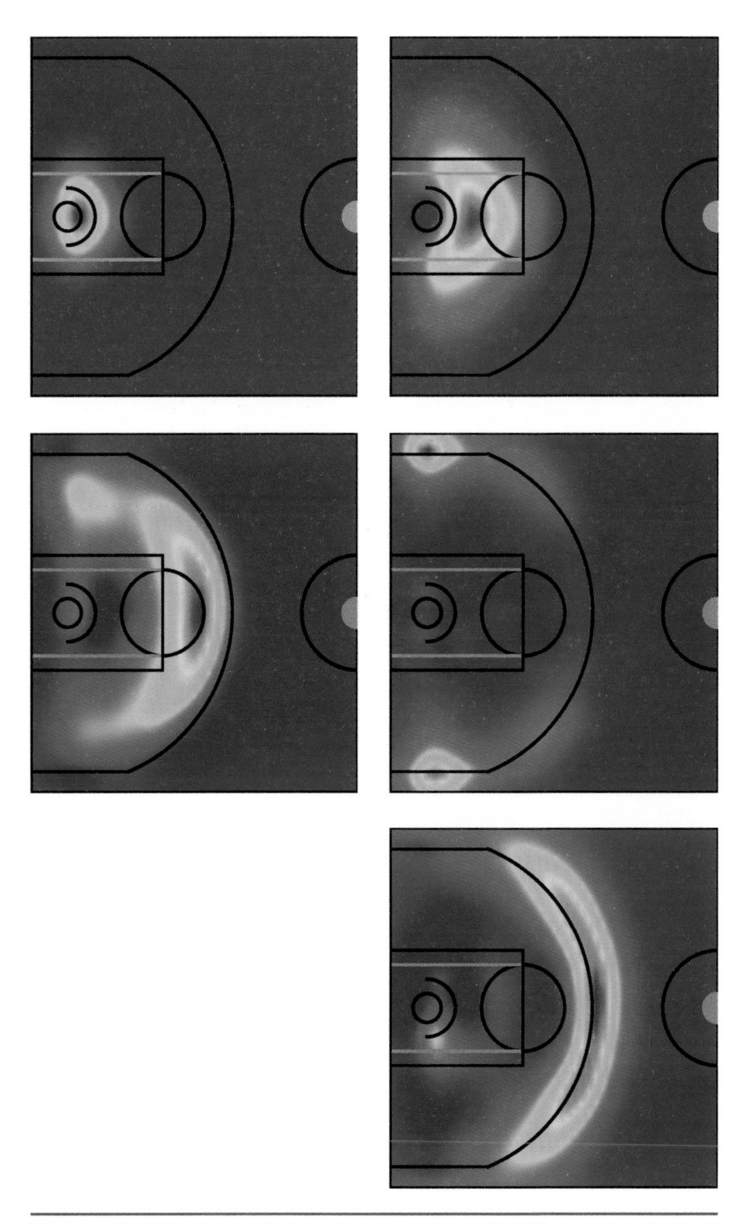

Figure 6. *"Heat map" showing location of different basketball shot types. Type 0, dunks (top left); type 1 (top right), layups; type 2, short jumpers (middle left); type 3, three-pointers from the corner (middle right); type 4, three-pointers from the arc (bottom right). (Figure courtesy of Institute of Mathematical Statistics, from "Characterizing the spatial structure of defensive skill in professional basketball," by Alexander Franks, Andrew Miller, Luke Bornn, and Kirk Goldsberry,* Annals of Applied Statistics, *2015, Vol. 9, No. 1, pp. 94–121.)*

The second matrix, **W**, tells you the types of shot that the offensive player likes to take. If there are five shot types, the columns would each contain 5 numbers. For instance, if player A takes 30 percent dunks, 40 percent layups, 30 percent midrange jumpers, and no 3-point shots, then the entries in his column

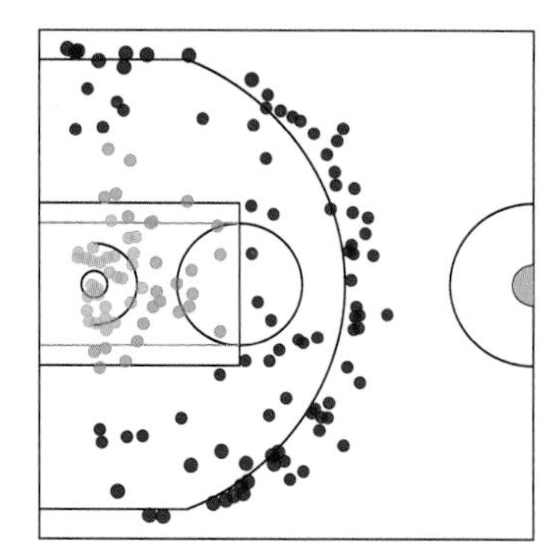

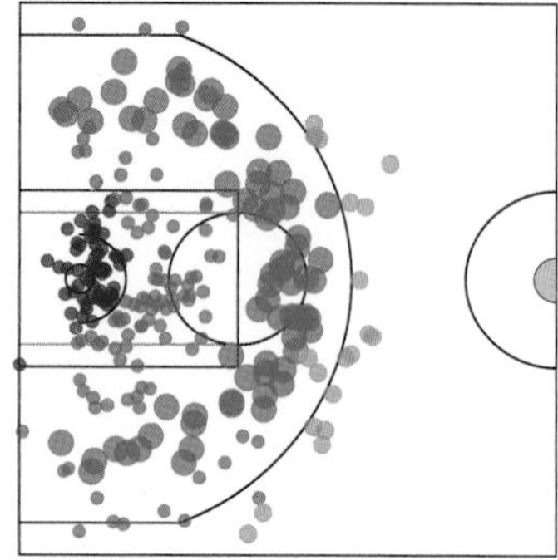

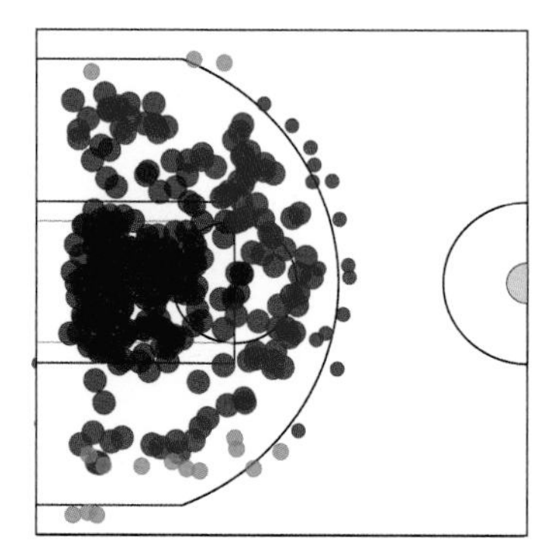

Figure 7. *Different kinds of defensive skill. Dots are coded by number of shots attempted (size) and percent made (color). Chris Paul (top left) is good at preventing shots from anywhere, and reduces his opponents' percentage made from the three-point arc. Dwight Howard (top right) prevents opponents from taking close-in shots, but allows a relatively high percentage of makes. Roy Hibbert (bottom) allows many shots close to the basket but is exceptionally good at stopping them from going in. (Figure courtesy of Institute of Mathematical Statistics, from "Characterizing the spatial structure of defensive skill in professional basketball," by Alexander Franks, Andrew Miller, Luke Bornn, and Kirk Goldsberry,* Annals of Applied Statistics, *2015, Vol. 9, No. 1, pp. 94–121.)*

would be (0.3, 0.4, 0.3, 0, 0). There would be one column for every player in the league, therefore 450 columns.

The idea of NMF, then, is to make the product, **LW**, as close as possible to the data vector Λ without overfitting the data, i.e., without postulating too many different shot types. The advantages of this approach are twofold. First, the huge 2350-by-450 data matrix is reduced to two more manageable matrices, one 2350-by-5 and the other 5-by-450. Second, it removes a lot of statistical "noise" from the data. Every dunk is drawn from the same probability distribution, no matter who took it and which precise location he took it from.

After this step, the XY Hoops team could identify which defender is marking which offensive player, and describe the shot-taking tendencies of each offensive player. The final step was to measure the efficiencies of the defenders against each different kind of shot. For this step, they used a technique called "*k*-means clustering," which separated the defenders into three different types and then inferred each player's contribution in comparison to players of the same type. This is a type of *hierarchical model*, commonly used in statistics to improve predictions of interventions. In the basketball case the hierarchical model was quite simple, with only two layers (defenders and defender types).

Putting all the ingredients together, XY Hoops could do something nobody had ever done before: estimate how many points a player was responsible for giving up in a given game or season. They also obtained some fascinating insights into defense (see Figure 7). Chris Paul, known as a great defensive player, reduced both his opponents' shot frequencies (indicated by the dot sizes in Figure 7, top left) and their percentages on outside shots. Roy Hibbert and Dwight Howard, known as excellent interior defenders, made an impact in dramatically different ways. Hibbert's team funnels shooters toward Hibbert, so he ends up defending lots of shots (large dots) but reducing their efficiency (dark blue). On the other hand, Howard is not as good as Hibbert at reducing shot efficiency near the basket (light blue), but he prevents such shots from ever being taken (small dots). "Kirk calls it the Dwight effect," says Franks.

"This kind of information is absolutely more useful for the teams than the fans," Franks says—although it's pretty interesting for fans, too. A team, for example, could compare its defenders' positions to the positions they were *supposed* to occupy, or it could identify its players' strengths and work to improve their weaknesses.

The Good, the Bad, and the Bogus

One big challenge remains for teams that want to dive into the world of sports analytics: How can they separate the good from the bad and the bogus? "A lot of teams don't know what they don't know," says Franks. "It's hard for them to evaluate who is legitimate and who is not." Says Lopez, "The easy way [for them] is to go with the person who sounds the most confident and has the prettiest pictures."

Research like that done by Baumer, Jensen and Matthews in baseball, and the XY Hoops group in basketball, is easy to pick out because it is published in research journals, conducted to professional standards, and uses true statistical modeling rather than mere spreadsheets of numbers. But these judgements can only be made because the papers are published openly. This suggests that no matter how hard teams try to preserve their secrets, they will still need academics, bloggers, and independent publications like *Baseball Prospectus* for quality control.

The good news, though, is that there is no shortage of interest among academics and the lay public. "My thesis was in genomics and proteomics," says Franks. "But my basketball work gets ten times the interest that the other research does. It's a way for mathematicians to connect with students and with other people, for sure."

The good news, though, is that there is no shortage of interest among academics and the lay public. "My thesis was in genomics and proteomics," says Franks. "But my basketball work gets ten times the interest that the other research does. It's a way for mathematicians to connect with students and with other people, for sure."